POWER BIBLE

BIBLE STORIES TO IMPART WISDOM

2

Moses, Leader of the Israelites

Green Egg Media

POWER BIBLE

BIBLE STORIES TO IMPART WISDOM

2

Moses, Leader of the Israelites

Green Egg Media

Publishing has been a dream that I've carried for a long time. The thought of being creative and expressing ideas is an incredible opportunity.

When we started Green Egg Media, our goal was to accomplish three things – to educate, to encourage, and to inspire. These objectives came from seeing the struggles and challenges that my children face today. It also came from reflecting on my childhood and seeing how my parents tried to educate, encourage, and inspire me. I love you, Mom & Dad!

In publishing our first book series, I can't think of anything that could educate, encourage, and inspire more than the Bible. The Bible provides wisdom in life, it lifts up discouraged hearts, and it sparks passion to live for the glory of Christ.

The Bible can be intimidating for young children because it has a lot of pages and contains deep theological concepts. The Power Bible was developed to help young kids learn basic biblical principles in a format that is fun and easy. In an age where television, internet, video games, YouTube, and Facebook all battle for our kids' time, we felt that this rich and illustrative format was important to grab and hold their attention. We want kids to develop a love for the Bible at an early age because it is critical to knowing and loving Christ.

II Timothy 3:16 – All Scripture is inspired by God and profitable for teaching, for reproof, for correction, for training in righteousness.

Romans 1:16 – For I am not ashamed of the gospel, for it is the power of God for salvation to everyone who believes...

Our team's desire is that the Power Bible will be a blessing to you and your family. We seek to honor Christ and hope that both kids and parents will come to know Christ as their Lord & Savior.

May Christ be exalted!

Gary J. Kim
President & CEO

The **B**ible can be a fun and amazing book.

The reason we have trouble sometimes is because the Bible was written more than 2,000 years ago. Also, while reading through its massive volume, people sometimes get tired. So we thought long and hard about how to make it easier for people to read. Today, we thank God for giving us this great and exciting story in the form of a comic book, so that people can read it from cover to cover.

We also worried about what would be the right dialogue to use so that the young readers of today would feel all the emotions from the stories of old. We did our best to show all the people and the events properly in comic form. We hope that you will become immersed in the Biblical world from the very first page, and we pray that you will learn the truths and wisdom of the Bible so that you can know God and His Love.

Artists Shin-joong Kim & Sook-ja Yum

Thank you to the publisher for making this children's comic Bible possible.

I realized a lot of good things working on this series. The comics and pictures are able to clearly convey the true content of the Bible to an elementary school student.

I'm sure that the young readers of this comic Bible series (all 10 books) will feel that the characters in the Power Bible are vibrant and alive, calling them deeper into the story and the journey.

Children learn from their environment. These days, children are flooded with information and images from numerous media. In the midst of this, we will continue to work with a higher vision and great patience in order to raise these children into truly faithful people.

Supervised by Pastor Jong-hyuk Kim

Table of Contents

Chapter 3: Toward the Promised Land

Cast of Characters

Moses

Leader of the Israelites. He was adopted by an Egyptian princess. God called him on Mount Horeb to lead the Israelites out of Egypt. He received the Laws and the Ten Commandments from God and gave them to Israel. He led them to the Promised Land, but suffered throughout the journey because the people were disobedient and rebellious. He got to see the Promised Land before he died, but did not enter it.

Aaron

Moses' brother and the first High Priest of Israel. He helped Moses lead the people of Israel out of Egypt. He continued to lead until his death, after which his son Eleazar became the High Priest.

Miriam

Moses' sister. When Moses was a baby, his mother put him in a basket in the Nile River, and Miriam watched out for him. She suggested her own mother to the Princess of Egypt as a nurse for Baby Moses. Miriam was the first prophetess of Israel, and a strong supporter of Moses.

Joshua

Moses' assistant and the next leader of Israel. He led the Israelites in battle against Amalek and many other nations while the Israelites wandered the wilderness. He became the leader of Israel once they reached the Promised Land. He challenged Israelites to serve God in sincerity and truth.

Pharaoh of Egypt

The leader of the Egyptians. He continued to enslave the people of Israel after his father. When Moses called him to let the Israelites go, he did not believe in the power of God. God eventually brought ten plagues to Egypt through Moses to convince Pharaoh to let the people of Israel go.

Balaam

The prophet son of Beor. He was asked by Balak, King of Moab, to curse Israel. Balaam was tempted by the money that Balak offered him and agreed to curse Israel. But, in the end, God did not allow him to curse Israel, and he blessed them instead.

The People of Israel

God's chosen people, the descendants of Abraham and Isaac through Jacob. They prospered in Egypt at first, but were later enslaved by Pharaoh. God used Moses to lead them to the Promised Land, but they constantly doubted God and complained. God punished them by making them wander in the wilderness just outside the Promised Land for forty years.

GOD SAW THE SONS OF ISRAEL, AND GOD
TOOK NOTICE OF THEM.
EXODUS 2:25 (NASB®)

Chapter
1

A Glorious Escape

GOSHEN

1. A Baby From the River
(Exodus 1:3 — Exodus 2:15)

Listen up! Pharaoh has spoken!
A state coffer is to be built here in Goshen!

State coffer? What's that?

It's a building that contains the Pharaoh's possessions.

But why build it here?

All Hebrews ages 20 and above, both male and female, shall work on the construction. From now on, you are all slaves of Pharaoh!

What? Slaves?

But we're the people of God!

This land was given by Pharaoh to our forefather, the great Minister Joseph, for saving Egypt from a 7-year famine!

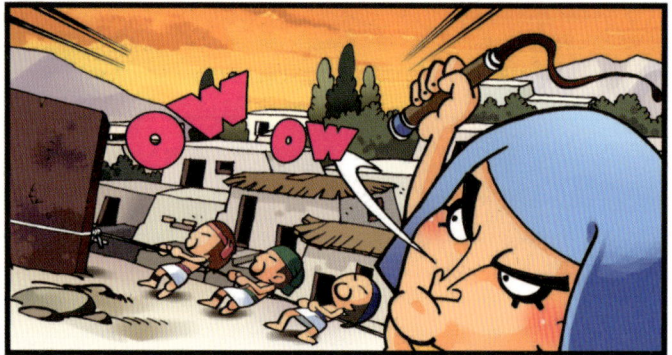

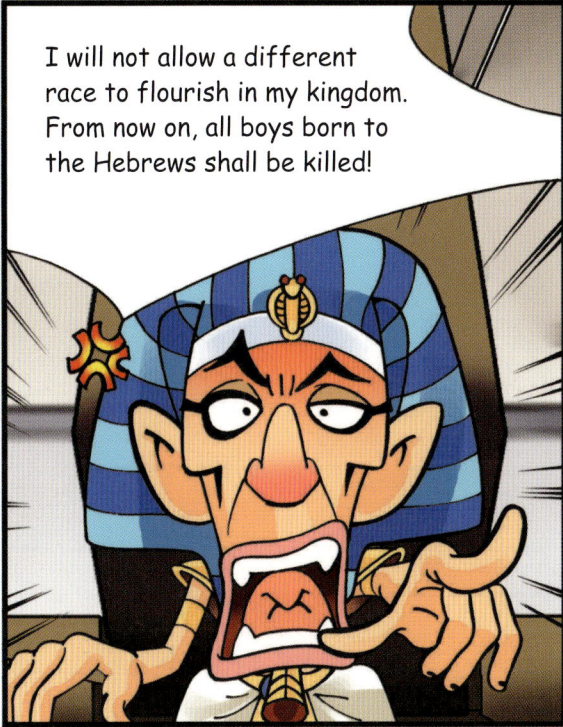

I will not allow a different race to flourish in my kingdom. From now on, all boys born to the Hebrews shall be killed!

Oh! A most excellent decision, great Pharaoh!

......

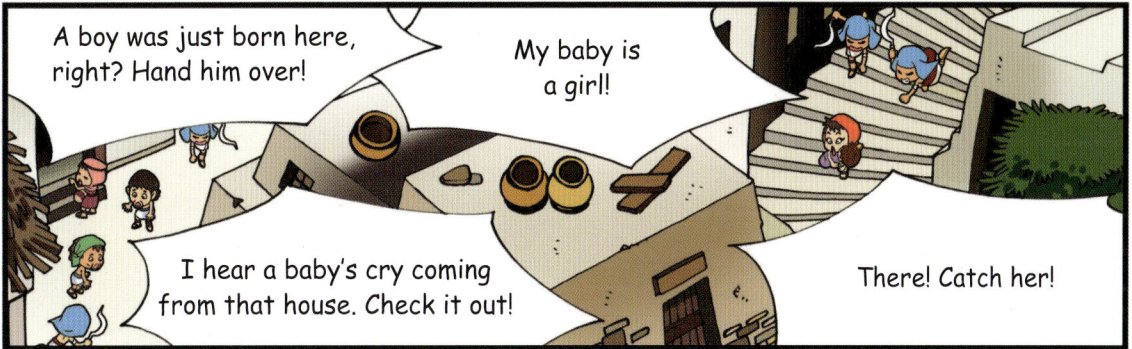

A boy was just born here, right? Hand him over!

My baby is a girl!

I hear a baby's cry coming from that house. Check it out!

There! Catch her!

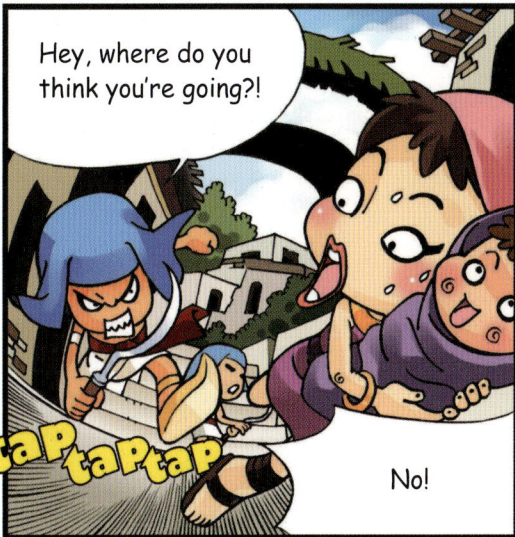

Hey, where do you think you're going?!

tap tap tap

No!

Aah! My baby!

15

Sob... My baby...

That's what you get for being born a Hebrew!

..!

Welcome home, father.

Hello, everyone. Is our baby still safe?

We can't hide him any longer. It's just a matter of time until they find out.

I know. That's why tomorrow at daybreak, I will place our baby in a basket and float him down the river. God will surely protect this child.

Miriam. Follow him and see where he goes.

Yes, mother.

Farewell, my son... May the Lord protect you...

Sob...

So refreshing! Everyone, come in the water.

Don't go too far, Princess.

float

Oh? What is that?

Look, a mysterious basket. Bring it to me.

Oh, it's a baby! So cute...!

He must be a Hebrew baby!

They must have sent him downstream in order to save him!

We cannot kill such an adorable child. Peek-a-boo!

Princess, that means...?

I will raise him as my own son! His name will be Moses, because I took him out of the water!

Princess! Shall I find a nurse for you?

Gasp!

A kid just appeared out of the water!

Splash

Yes, if you could. I will give her wages...

AFTER SEVERAL YEARS, THE PRINCESS OF EGYPT HAD MOSES BEGIN HIS EDUCATION AS A PRINCE OF THE PHARAOH.

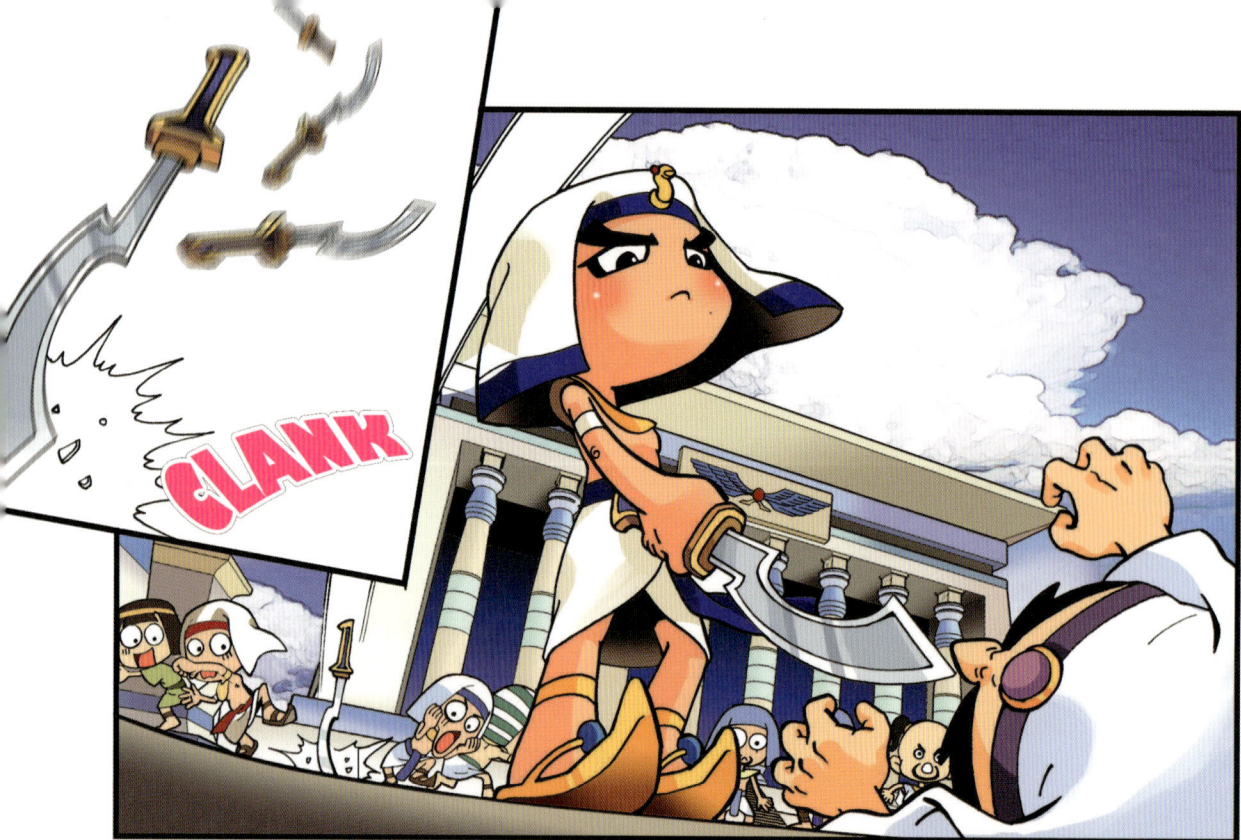

CLANK

Amazing, Moses!

clap
clap

Mother...!

Princess...

You defeated the training captain who is the greatest swordsman in the land.

I still have much to learn.

On the contrary. The Prince's swordsmanship is extraordinary.

I am very proud of you. You are truly worthy of becoming king.

I am honored.

But how can a Hebrew become the Pharaoh of Egypt?

I wish to go visit Goshen.

Certainly. A good king should see how the people live, and he should learn how to deal with slaves...

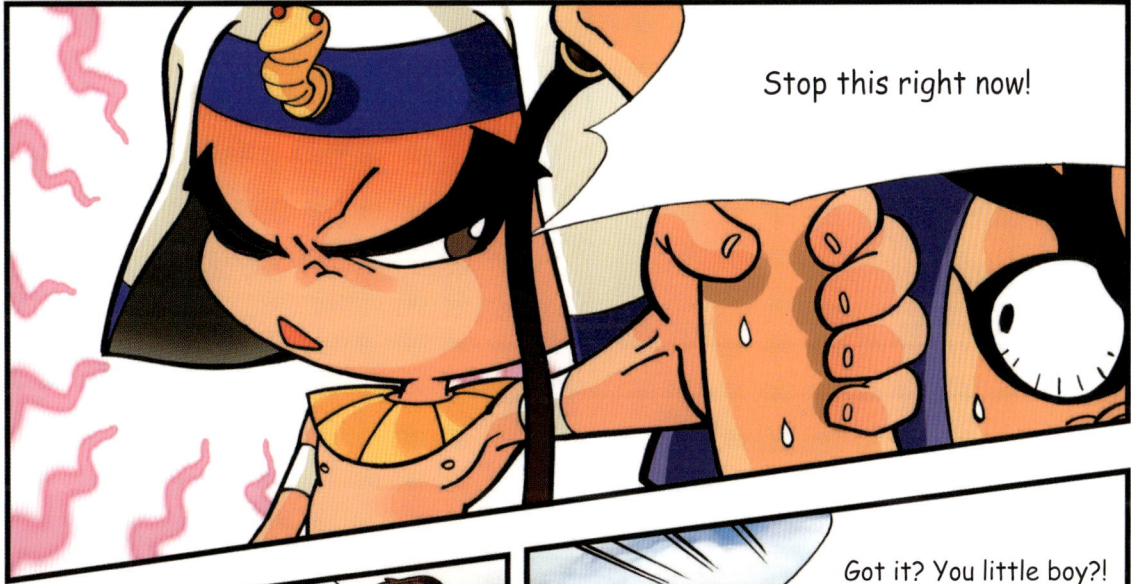

SMASH

crash

He's... dead!

It was definitely Prince Moses!

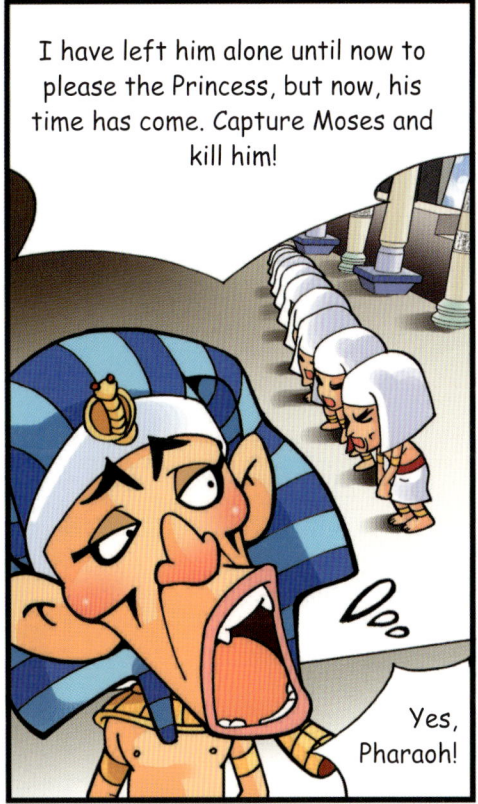

I have left him alone until now to please the Princess, but now, his time has come. Capture Moses and kill him!

Yes, Pharaoh!

THUMP

2. A Burning Bush
(Exodus 2:16 – Exodus 4:31)

WHEN IT WAS DISCOVERED THAT HE HAD KILLED A MAN, MOSES FLED AND ESCAPED TO THE MIDIAN WILDERNESS.

Canaan

Philistia

Kadesh-barnea

Raamses

Succoth

Baal-zephon

Midia

Elath

Marah

Elim

Red Sea

Mount Sinai

huff

huff

That's...!

Water...

Water...

Water!

slap

slap

slap

gulp

gulp

Oh, I can live again!

thump

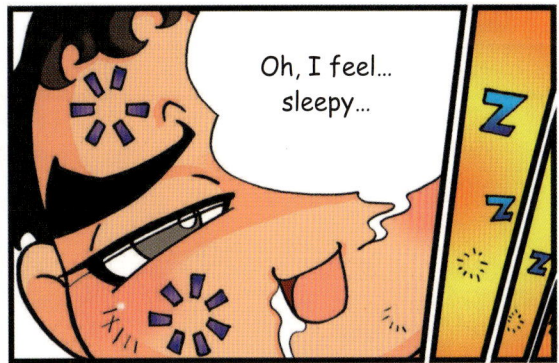

Oh, I feel... sleepy...

Who is that?

He doesn't look like he's from around here. A traveler?

His clothes are all tattered, but he's really handsome.

Oh, he woke up!

What, a dream...?

Hee-hee...

Ho ho~

Hee-hee.

splash

Hey! The well is free. Line up the sheep!

baa

Hey! We were here first!

That's right! Our sheep were lining up to drink the water!

Hey, line them up!

Line up!

You might have come here first, but you were just fooling around and not using the well. So our flock will drink first.

Who was fooling around?!

Oh, these little girls...

......!

A beggar, you say...?

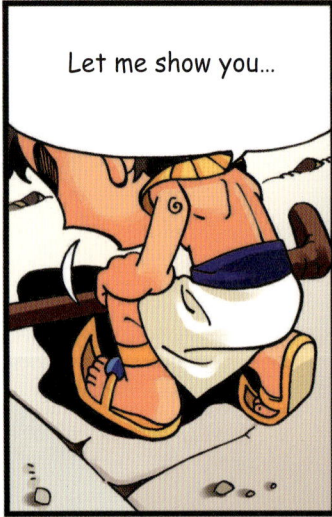

Let me show you...

Just how this beggar named Moses deals with people like you!

What is he talking about?

His eyes are on fire.

Is he really challenging all of us to fight?

Should he be saying things like that?

I don't know, but he sure sounds cool!

Hoo... Alright, so you want to fight. I know you're trying to look cool in front of the girls.

Oh...

OK, we'll take you on.

We'll give you a taste of a Midianite staff!

You'll regret showing off!

Swish

Swish

THE BEAUTIFUL SEVEN SISTERS WERE THE DAUGHTERS OF REUEL, THE MIDIANITE PRIEST. MOSES MARRIED THE ELDEST NAMED ZIPPORAH, AND HE WENT TO LIVE IN THE HOUSE OF REUEL.

MOSES TENDED HIS FATHER-IN-LAW'S SHEEP. HIS WIFE ZIPPORAH GAVE BIRTH TO A SON, AND THEY LIVED HAPPILY.

Got you! What a brave lamb. But what would you do if a wolf came along?

pause

What's that? A light...?

WHRRR

The bush is on fire...

But... How can it be...? It isn't being destroyed by the flames...!

I'd better go closer and check it out!

Moses...

Do not come near here.

Remove your sandals because you are standing on holy ground.

I am the God of your father, the God of Abraham, the God of Isaac, and the God of Jacob.

The God of my ancestors...?

Me...?

Here I am...

I have met God...

I am a dead man now...!

kneel

I have seen the suffering of My people who are in Egypt, and hear their cry. So I have come down to deliver them from the power of the Egyptians, and to bring them up out of Egypt to a land flowing with milk and honey.

You will do that work.

What do You mean?

Moses. I will send you to Pharaoh, so that you may bring My people out of Egypt.

...!

Lord... Who am I that I should do such a thing? I was driven out of Egypt by Pharaoh.

Do not fear, for I shall be with you.

WHRRR

But, Lord... I don't even know Your name.

When the people of Israel ask, "What is His name?" What shall I say?

I am Who I am, the Lord*.

Thus you shall say to the sons of Israel, "The Lord has sent me to you."

The Lord...

This is My name forever, and this is the name I will be remembered by for all generations.

WHRRR

You will go with the elders of Israel to the King of Egypt and you will say to him, "The Lord, the God of the Hebrews, has met with us. So now, please let us go a three days' journey into the wilderness, that we may sacrifice to the Lord our God."
What if they don't believe me?
I can't really believe it either.
What do you have in your hand?
A staff.
Throw it on the ground.
What do You intend to do?
tap
HISS
It turned into a snake...!

Grasp the snake by its tail.

Now, do you believe? I will show you one more thing. Put your hand on your chest.

My chest...?

Ack! It became leprous!

GASP

Now, put it on your chest once more.

Yes!

It's clean again...! Oh, I almost had a heart attack.

They will believe after seeing those two signs.

But if they do not believe, then you shall take some water from the Nile and pour it on the dry ground; and the water that you take from the Nile will become blood.

Lord...! I have never been good at making speeches.

Who has made man's mouth? Is it not I, the Lord? So go, and I will be with you and teach you what you are to say!

WHRRR

Oh, Lord! Please send someone else.

Your brother Aaron, the Levite, speaks well! He is coming out to meet you!

I've run out of excuses...

I will obey the Lord.

Moses!

Aaron!

3. Let My People Go
(Exodus 5:1 – Exodus 7:13)

IT HAD BEEN 40 YEARS SINCE MOSES HAD LEFT EGYPT, AND A NEW PHARAOH HAD COME TO THE THRONE.

Hello, Moses. I remember studying together in our youth. You were skilled with the sword.

When father passed away, your crime of murder was erased. But I never thought you would come back so boldly. Well, what do you want?

The Lord God of the Hebrews has met with us.

Please, let us go a three days' journey into the wilderness so we may sacrifice to the Lord our God!

Humph.

Ha Ha Ha

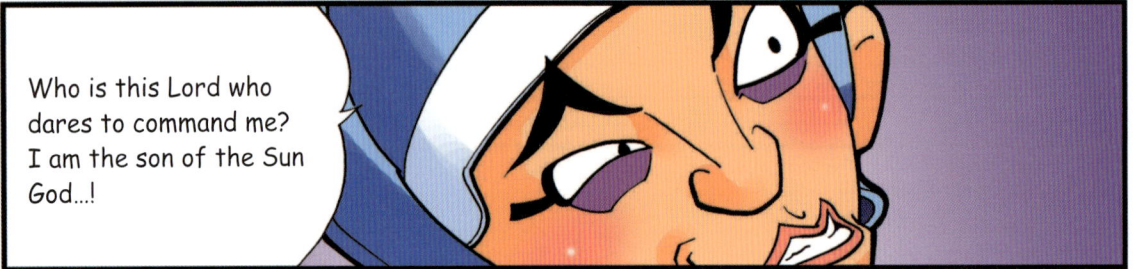

Who is this Lord who dares to command me? I am the son of the Sun God...!

Why should I listen to you and let the people of Israel go?

He is the Lord, the God of our ancestors. Let my people go. If you do not, terrible things will happen in Egypt.

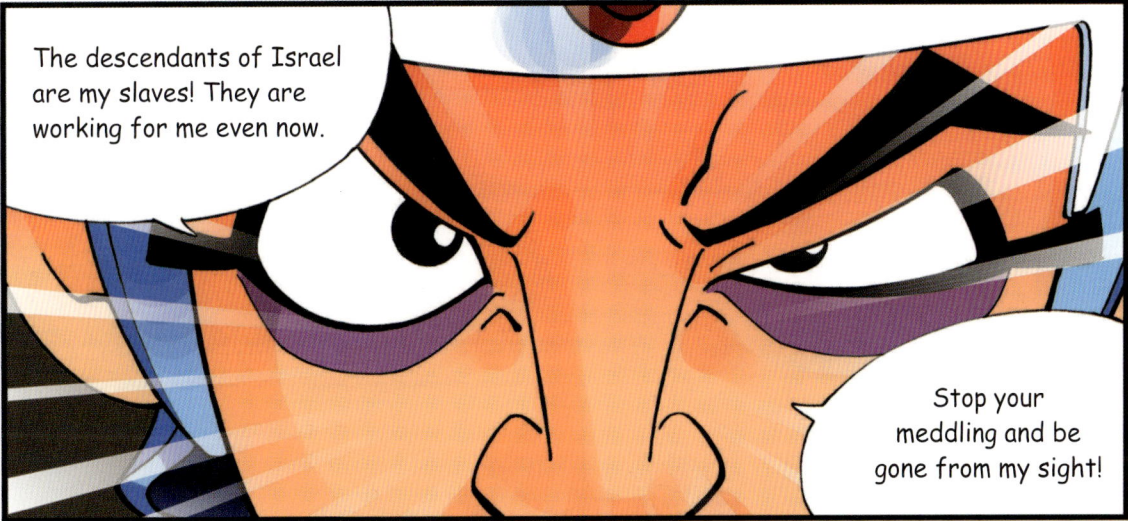

The descendants of Israel are my slaves! They are working for me even now.

Stop your meddling and be gone from my sight!

Hear me! From now on, the Hebrews will not be given any straw to make the bricks. They will gather the straw themselves. But the number of bricks required shall remain the same!

Hmm, they must have some spare time on their hands if they can come up with this silly request for a sacrifice.

The required amount of bricks must be filled!

gasp

No way... We barely can do it now... How can we keep it up if we have to gather the straw ourselves...?!

Understand?!

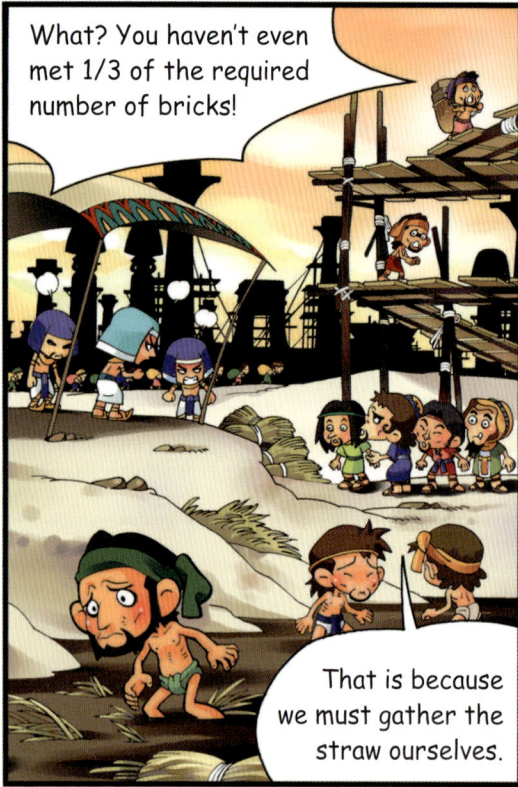

What? You haven't even met 1/3 of the required number of bricks!

That is because we must gather the straw ourselves.

Well, move faster!

We can't keep up with this!

Let's go beg the Pharaoh.

Do you think he will listen?

We were barely able to meet the previous number of bricks. Now, if we must gather the straw ourselves, there is no way that we can meet the requirement. We beg that you reconsider...

You lazy scoundrels... You cannot even meet this amount, yet you ask for time to go make a sacrifice? I will not provide the straw, and the amount will remain the same. You must take responsibility for this yourselves!

This is all the fault of Moses and Aaron!

Humph! Speaking of which...

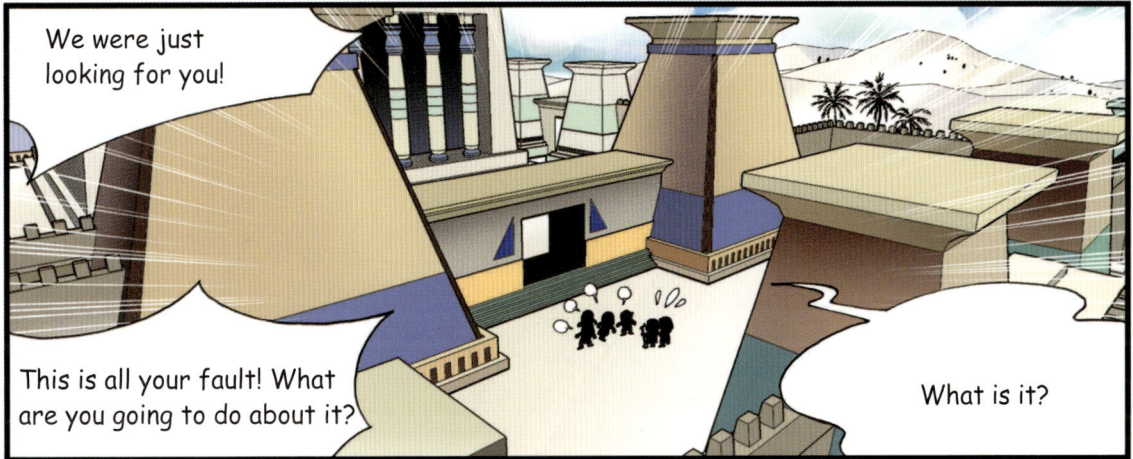

We were just looking for you!

This is all your fault! What are you going to do about it?

What is it?

You said that you would save us from slavery, but you just made things worse!

Are you planning to kill us completely?

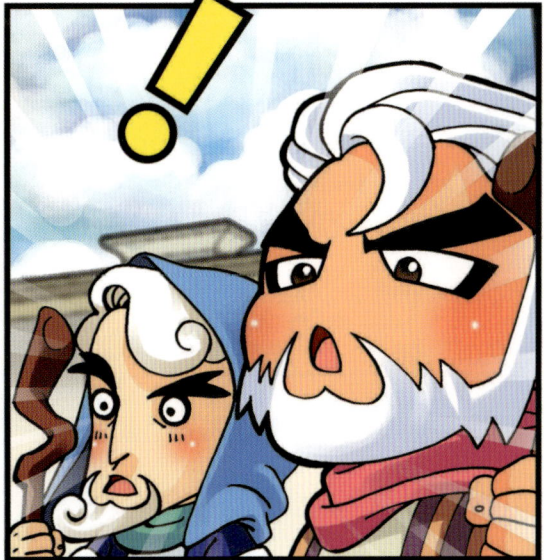

O Lord! Why have You sent me here? I only did as You instructed, but now my people are suffering even more!

I haven't saved them. I have made things worse!

Now you shall see what I will do to Pharaoh.

Shock

I am the Lord. I will deliver you from the rule of the Egyptians and bring you to the land that I swore to give to your forefathers. I will redeem you with an outstretched arm and with great judgments. Then, you shall know that I am the Lord your God.

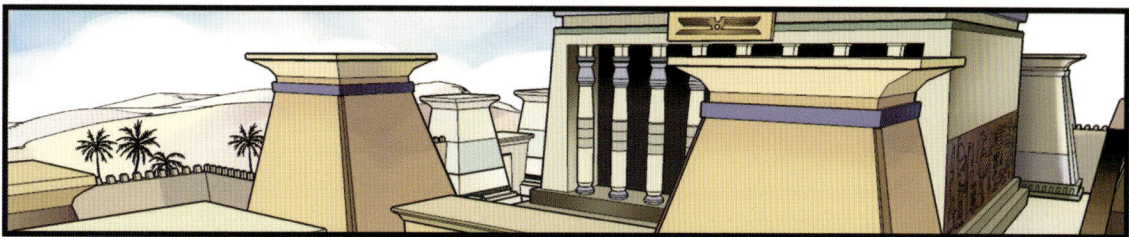

48

4. The Ten Plagues
(Exodus 7:14 – Exodus 12:30)

GOD HARDENED THE HEART OF PHARAOH, AND HE REFUSED TO LISTEN TO MOSES.

Why do I have to see your angry faces so early in the morning?

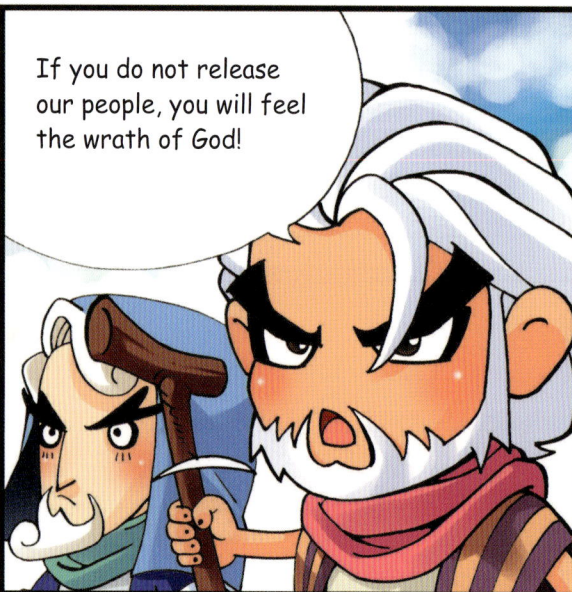

If you do not release our people, you will feel the wrath of God!

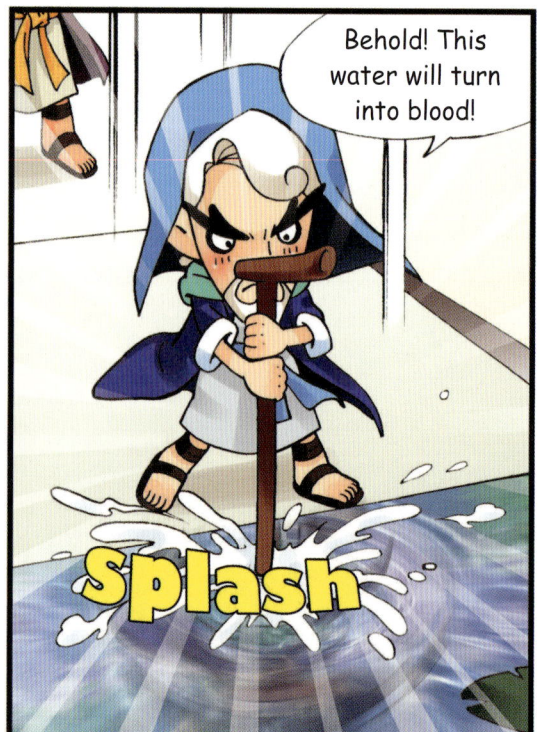

Behold! This water will turn into blood!

Splash

You say that too often. OK, I'm looking. What will you do?

SHHH

snap

Abracadabra...

SHHH

Brother!

Right!

51

Let all waters of the lakes, reservoirs, and pools in Egypt turn to blood!

Oh, yeah? Let all waters throughout Egypt turn to blood!

We won't lose!

Those guys...

Oh, no! The jar is full of blood!

Aaah

Even the well is full of blood!

The Nile River has turned to blood, too!

Oh, this smell...! There is a terrible stink everywhere!

THE NILE RIVER WAS FULL OF BLOOD FOR 7 DAYS. HOWEVER, PHARAOH'S HEART WAS UNCHANGED.

Let frogs swarm over the land of Egypt!

If you don't change your mind, you will regret this!

I don't know the meaning of the word "regret*."

52

* REGRET: to be sorry for

Alright! Get rid of the frogs! Then, I will let your people go so that they can sacrifice to the Lord!

You should have said that from the beginning. I will ask God to send the frogs back to the Nile River tomorrow.

BUT WHEN THE FROGS DISAPPEARED, PHARAOH WENT BACK ON HIS PROMISE.

Foolish Pharaoh. You still don't know what is in store for you!

thud

Do you think I'm crazy? Why would I let my property go?!

What's this?!

SWARM

Gnats*! They're everywhere!

Yikes, someone help!

Oh, so itchy...!

54

*GNATS: tiny flying bugs that bite

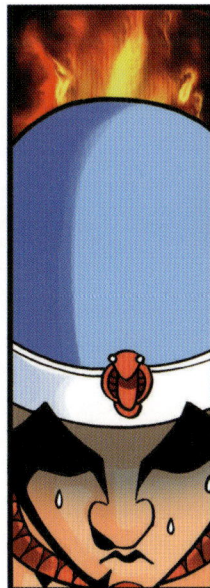

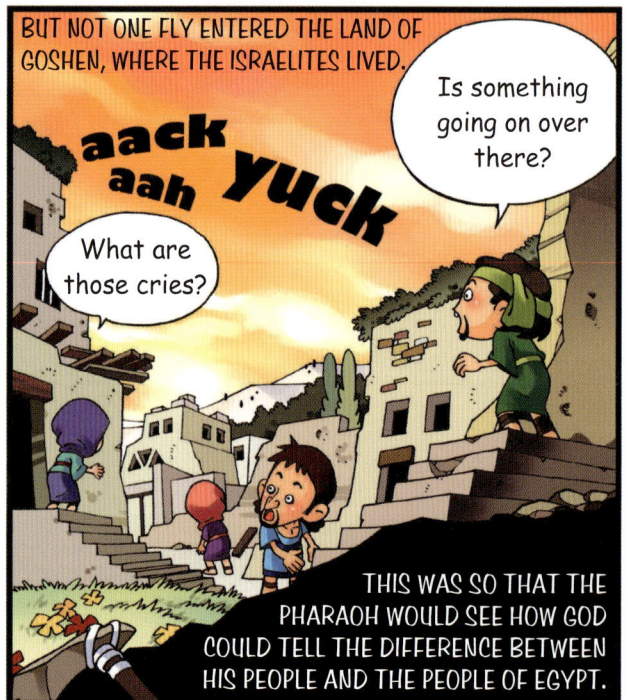

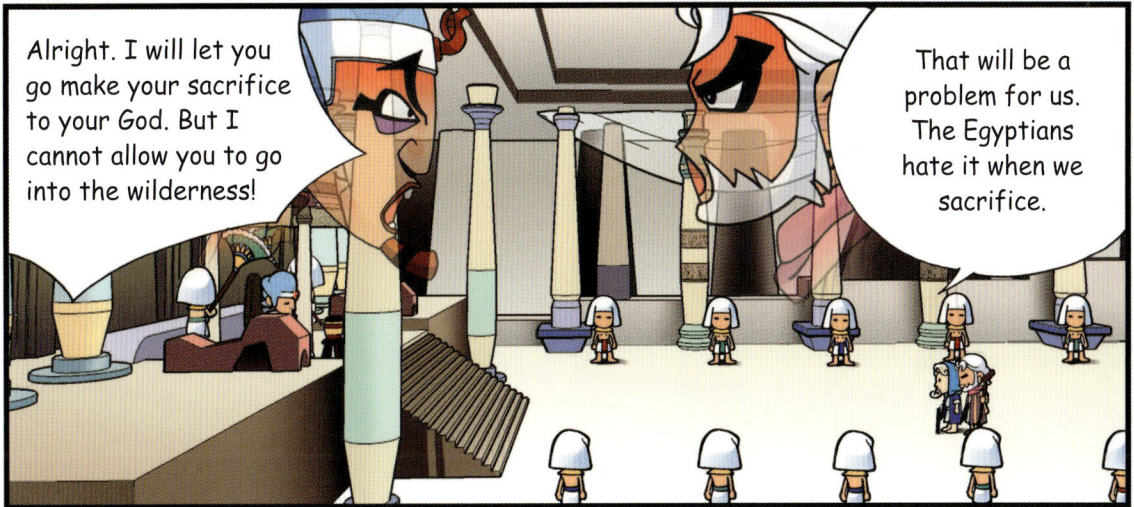

GOD CONTINUED TO BRING PLAGUES UPON THE LAND OF EGYPT.
All of our cattle died in the middle of the night!
So did our sheep, ducks, and camels!
It must be because the gods Hathor and Khnum didn't like our offerings.
No, I heard it's the curse of the Hebrew God.
I heard that a cow disease has broken out in Egypt.
And many other animal sicknesses!
We're safe. God will protect us.
Behold! This land's people and animals will be covered in boils and sores!
Go ahead! See if I even bat an eye!
Let hail fall from the sky!
RUMBLE
Watch out!
It's hail that is the size of a man's fist!

Locusts!

They're swarming like a cloud!

Everyone, flee!

BUZZ

No! We barely saved this food!

Great Pharaoh, your people are full of grief and bitterness... So...

Fine! I will let the Hebrews go!

Who will go to make the sacrifice?

All of the men, women, and children of Israel, along with their cattle.

No! Only the men may go! I won't negotiate with you any further. Get out!

SLAM

This is serious... The Hebrew God is certainly stronger than our gods...

The blood in the Nile was a challenge to Osiris...

The frogs, to Heqet, the flies, to Uazit...

The cattle disease, to Hathor and Khnum, the boils, to Amunhotep and Sekhmet,

The hail, to Shu, and the locusts, to Serapis...

And now Egypt's greatest god Ra has been defeated...

Then, Pharaoh, the son of the Sun God...

Is dying of embarrassment!

grrr

That is all I have to say, so stop appearing before me! The next time you appear, you will be killed!

Alright, this is my final decision: You may take your wives and children, but you cannot take your livestock!

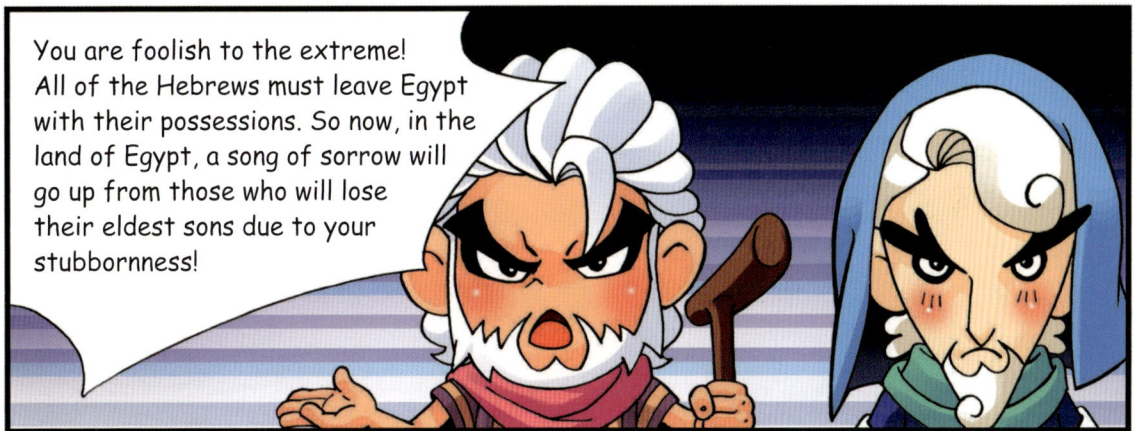

You are foolish to the extreme! All of the Hebrews must leave Egypt with their possessions. So now, in the land of Egypt, a song of sorrow will go up from those who will lose their eldest sons due to your stubbornness!

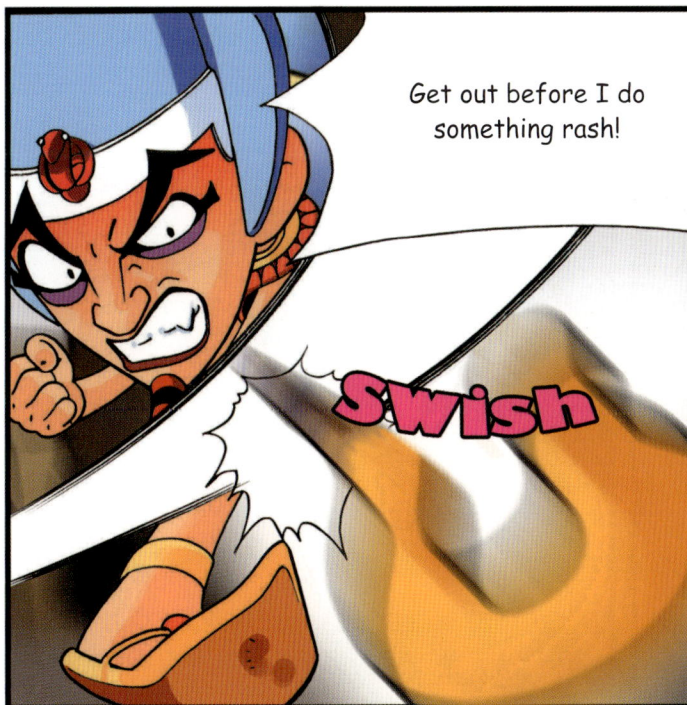

Get out before I do something rash!

Swish

turn

I'm leaving! Do you think I enjoy being here?

Everyone! I will deliver God's message to you. You must follow every single instruction.

The Lord will kill all firstborn males of both people and animals in Egypt. Then, Pharaoh will no longer be able to refuse.

You must kill a young male from your flock and keep the Passover.

Why do we have to eat unleavened bread and bitter herbs?

That's what God commanded. This is to keep the Passover.

Why are you putting lamb blood around our door?

When God comes to kill the firstborns of Egypt, He will see this blood and pass over our house.

Passover means that the Lord's judgment passed over us and spared the houses of the Hebrews.

You will eat all of the roasted lamb meat and not have any leftovers. And you must eat it together with unleavened bread and bitter herbs.

Why bitter herbs?

So that we may remember the bitterness of 400 years of slavery in Egypt.

slam

Moses!

Oh, no!

No! My son!

Our firstborn calf has died!

It's begun!

huff

huff

My son... son of the Sun...

boom

I've lost...!

I, the son of the Sun...
As well as all of the
gods of Egypt....
We have all lost to the Lord,
the God of our slaves...!

Go!

I never want to see any of you Hebrews again. Take what you wish and get out!

5. The Parting of the Sea
(Exodus 12:31 – Exodus 15:21)

Wow! We're free!

God saved us from slavery!

He kept His promise to our ancestors!

Hurrah for God!

Hurrah for Moses!

Your God is really amazing. Here, use this to make your statues of Him.

We'll give you animals for your sacrifices, too!

Oh, thank you!

So they finally realized it, huh? Our God is the strongest in the world!

Umm...

What is it? You're not Israelites like us...

We want to worship your God, too. This isn't our home anyway... Let us go with you.

Well... OK. We understand your wish to leave.

AFTER FOLLOWING JACOB TO EGYPT AND ENDURING 430 YEARS OF SLAVERY, THE ISRAELITES NOW BEGAN A LONG JOURNEY TO THE PROMISED LAND THAT GOD HAD PREPARED FOR THEM.

AT THE TIME OF THE ISRAELITES' DEPARTURE, THERE WERE ABOUT SIX HUNDRED THOUSAND MEN ON FOOT, NOT INCLUDING THE WOMEN AND CHILDREN. MANY OTHER PEOPLE WENT WITH THEM, AS WELL AS A LARGE NUMBER OF LIVESTOCK, BOTH FLOCKS AND HERDS.

THE ISRAELITES TRAVELED ACCORDING TO THEIR TRIBES.

?

clunk clunk

What's that?

Those are the bones of Joseph, our great ancestor. He instructed us to take his body with us when we left for the Promised Land.

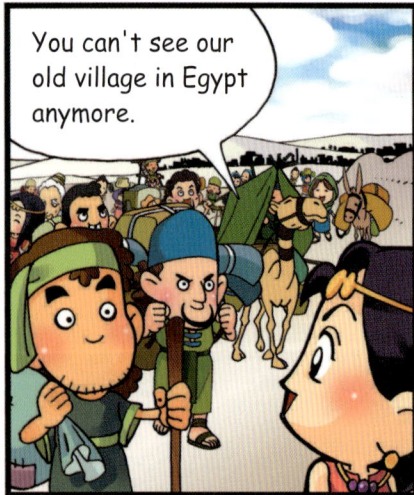

You can't see our old village in Egypt anymore.

There's nothing but wilderness ahead. I'm kind of scared...

What's to worry about when Moses is our leader?

Oh! Look at that?!

Whoosh

Everyone! God Himself will lead us with a column of cloud by day and a column of fire by night!

Wow...! God is really amazing!

Hallelujah!

What was I, crazy? How could I let all those slaves go?!

Who will complete our construction projects? And who will harvest our farms...?

O Pharaoh, why don't we just... bring them back?

What...?

We have been tracking them ever since they left.

poke

They are at Baal-zephon right about now.

Baal-zephon...? Isn't that right by the sea?

Yes, O Pharaoh! That foolish Moses must have gotten lost.

Prepare all of the troops and chariots! We will kill Moses and bring back all of the slaves and possessions!

What? It's the Red Sea!

What do we do now? Are we going to cross the sea on boats?

What boats?

Moses! Moses! Trouble!

slap slap slap

Everyone, calm down!

Do not be afraid. Wait and see what God will do! I promise that you will never lay your eyes upon Pharaoh or his army again!

So is this the end?

God performed all of those miracles. Surely, He won't let us die here.

Whoosh

Look! The cloud is moving!

S-s-s-s

It's going toward the Egyptian army!

God is going to fight them directly!

What's this?

That cloud came here, and suddenly it turned pitch-black!

Don't push!

Who's pushing?

Grrr...

It's the Lord again!

God's cloud is blocking the Egyptian army!

Not even an ant can get through!

Ah...!!

Oh, Lord...!

That's...

SH - SH

The sea just split open, right?

I've never even heard of something like this before!

Hurry! We must cross before the waters return!

Amazing!

The waters are like walls!

We're walking through the sea! How is this possible?

Look! I see fish!

That's the power of our God!

Hurry! Pharaoh's army is right behind us!

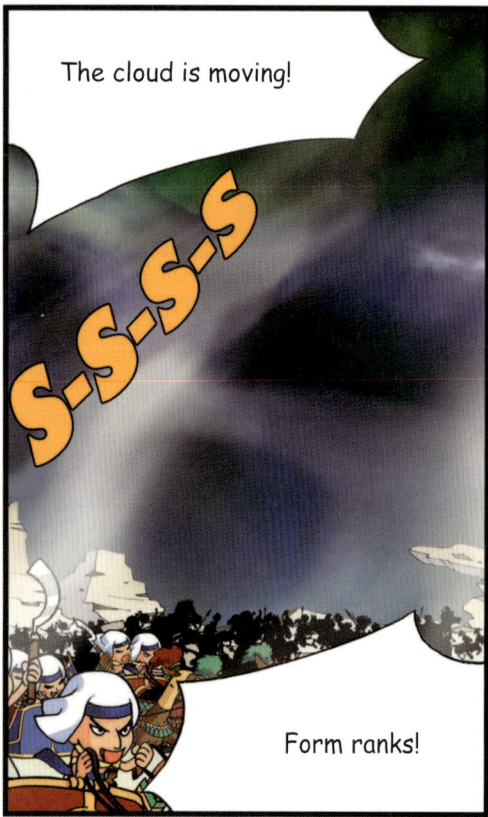

The cloud is moving!

S-S-S-S

Form ranks!

O Pharaoh, look!

What...!

After them!

Capture them all!

I can't believe it. They walked through the sea...!

Their God must be really powerful.

There is no way that we can defeat Him!

I said, after them! If you don't, I will kill you myself!

Oh...

tremble

Hyah!

rumble

We're dead either way! So let's just obey our Pharaoh!

The Egyptian army is coming!
Aaah
Hurry! Get to the other side!
We made it!
This way!
We're back on solid ground! Keep running!
rumble
Moses! What do we do?!
Just wait a little longer!

Chapter
2

God as the Guide

"IN YOUR LOVING KINDNESS, YOU HAVE
LED THE PEOPLE WHOM YOU HAVE
REDEEMED; IN YOUR STRENGTH, YOU
HAVE GUIDED THEM TO YOUR HOLY
HABITATION." EXODUS 15:13 (NASB®)

6. God Feeds His People
   (Exodus 15:22 – Exodus 18:27)

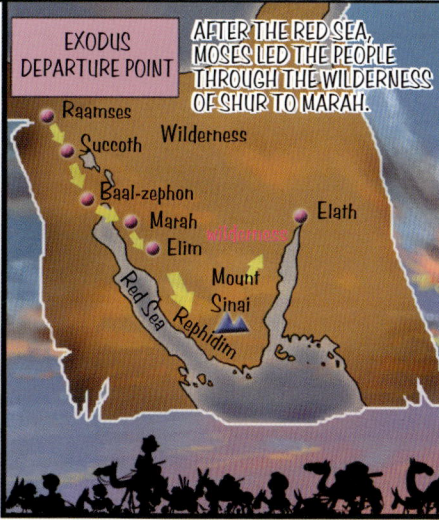

EXODUS DEPARTURE POINT

AFTER THE RED SEA, MOSES LED THE PEOPLE THROUGH THE WILDERNESS OF SHUR TO MARAH.

Raamses
Succoth
Wilderness
Baal-zephon
Marah
Elim
Elath
Mount Sinai
Red Sea
Rephidim

I'm parched... We ran out of water.

My feet hurt. I can't walk anymore.

We've been in the wilderness for 3 days, and we haven't had anything to drink.

huff

huff

Look! Water!

Wa... Water?

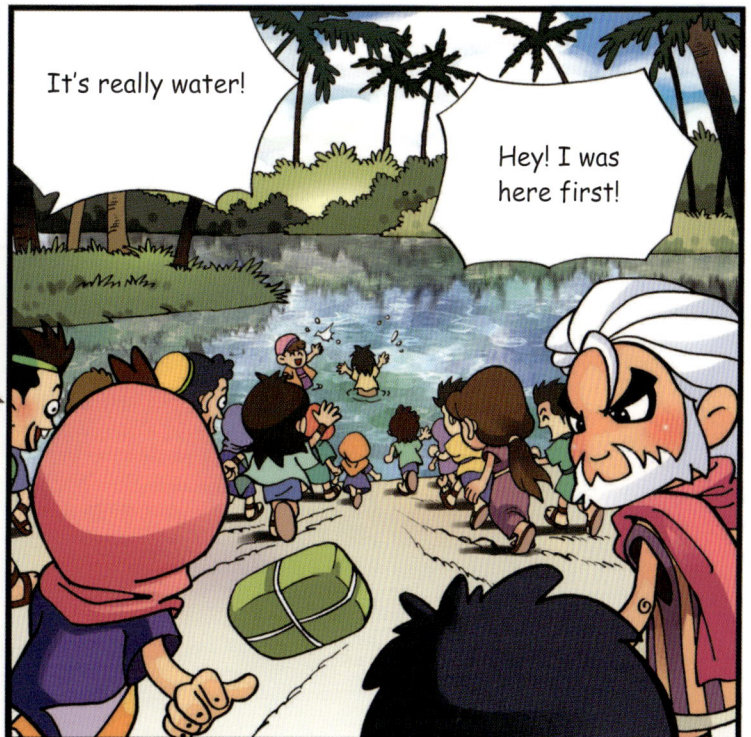

It's really water!

Hey! I was here first!

Amazing...! It's really sweet!

Real water!

Sweet and refreshing!

Let's fill all of the water jugs.

God is so powerful!

Listen, everyone.

If we obey the Lord and follow His commandments, God will not allow us to get the diseases that He gave to the Egyptians!

Yes, sir!

Just say the word! We will obey everything!

THEY NEXT ARRIVED AT A PLACE CALLED ELIM.
IT HAD 70 DATE PALMS AND 12 SPRINGS OF WATER.

It would be nice to live around here.

Don't say that. The land that God promised to us is many times better than this place.

WILDERNESS BETWEEN ELIM AND MOUNT SINAI. ONE MONTH AFTER LEAVING EGYPT.

huff

huff

huff

Father!

thump

Moses, what will you do?

We are out of food and water!

It would have been better to die of disease in Egypt than to starve to death in this barren wilderness!

That's right! At least we had bread and meat to eat in Egypt!

Unbelievable. Are you saying that it was better when you were slaves?

Better than starving to death!

You can't eat freedom!

Foolish people...

...

Everyone! When you complain to us, you are complaining to God. Are you aware of that?

The Lord saved you from Egypt with a mighty hand. Do you think He would now let you starve to death?

Now, everyone step forward!

Oh, look...!

SH-H-H-H

It's the Lord!

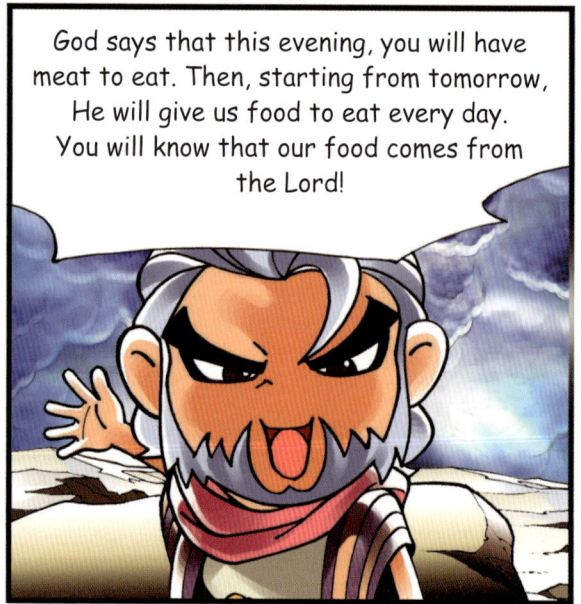

God says that this evening, you will have meat to eat. Then, starting from tomorrow, He will give us food to eat every day. You will know that our food comes from the Lord!

Something is flying this way...

They are quail birds!

Birds!

Catch them! They are meat!

Oh, thank You, God! You really sent us meat!

Mmm~ Smells good...

How long has it been since we had meat?!

Whrrr

THE NEXT DAY, SOMETHING STRANGE APPEARED ON THE GROUND.

Mom, there are strange white things on the ground.

It isn't dew.

And it isn't frost.

This stuff is everywhere!

What is it?

It's sweet!

THE ISRAELITES CALLED IT MANNA*.

*MANNA: a white food that tastes like wafers with honey

This is the food that God has provided. We will eat this every day. In the morning, gather enough for your family to eat.

You must gather only what you will eat that day. But on the 6th day, gather enough for the following day as well.

Who knows if this will be here again tomorrow? I'm collecting as much as I can and saving it.

Oh, I'm so full... I can't eat any more.

THE NEXT DAY

Oh, it's all rotten!

Amazing. The food from yesterday is still fresh.

That's because today is the Sabbath. We are not supposed to work, so God told us to collect our food yesterday.

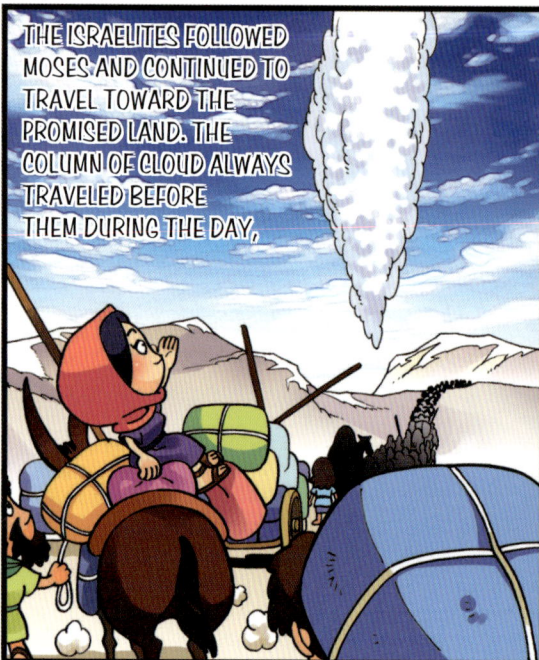

THE ISRAELITES FOLLOWED MOSES AND CONTINUED TO TRAVEL TOWARD THE PROMISED LAND. THE COLUMN OF CLOUD ALWAYS TRAVELED BEFORE THEM DURING THE DAY,

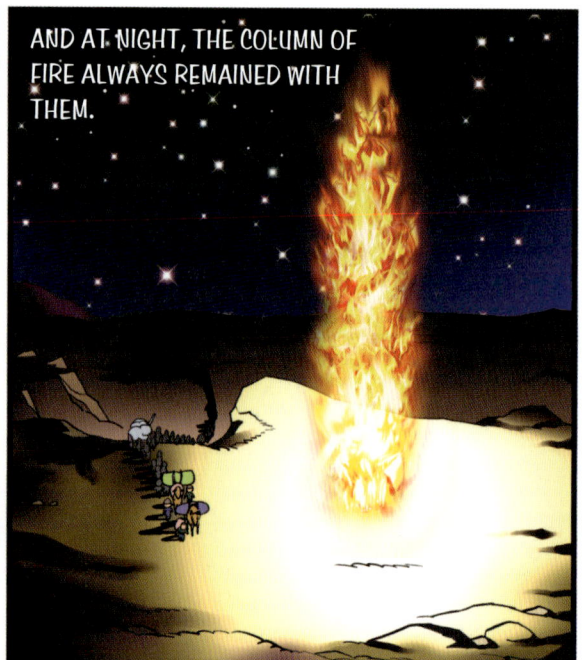

AND AT NIGHT, THE COLUMN OF FIRE ALWAYS REMAINED WITH THEM.

REPHIDIM

I can't go on!

Me, either!

splat

I'm dying of thirst!

Give us water or give us death!

Why did you bring us out of Egypt?!

We don't care about the Promised Land. Just give us water!

The Lord has abandoned us!

All they do is complain.

God has provided for them time and time again, but they forget about that and only think about their current discomfort.

Moses! Where are you going?

plod

plod

Lord...

I said from the beginning that I couldn't do this! So why...?

tremble

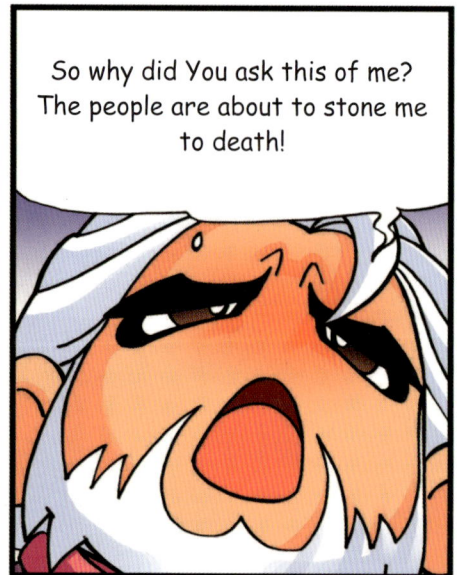

So why did You ask this of me? The people are about to stone me to death!

Rise, Moses. Take your staff and bring everyone to the rock on Horeb.

I shall stand there before you.

Everyone. See with your own eyes whether the Lord is among us!

crash

SH-H-H-H

Water!

Hallelujah!

The Lord is amazing!

Sorry for complaining, Moses.

It's alright.

Moses! Moses! The Amalekites are attacking!

Swish

What? Amalekites?

Their movements looked a little strange...

huff

huff

Joshua! Choose some men and go fight! I will stand on top of the hill with the staff of God in my hand!

Yes!

Let us help...

We will hold up your arms on either side...

Thank you...

Look! We are winning again!

RAAH

God must be helping us!

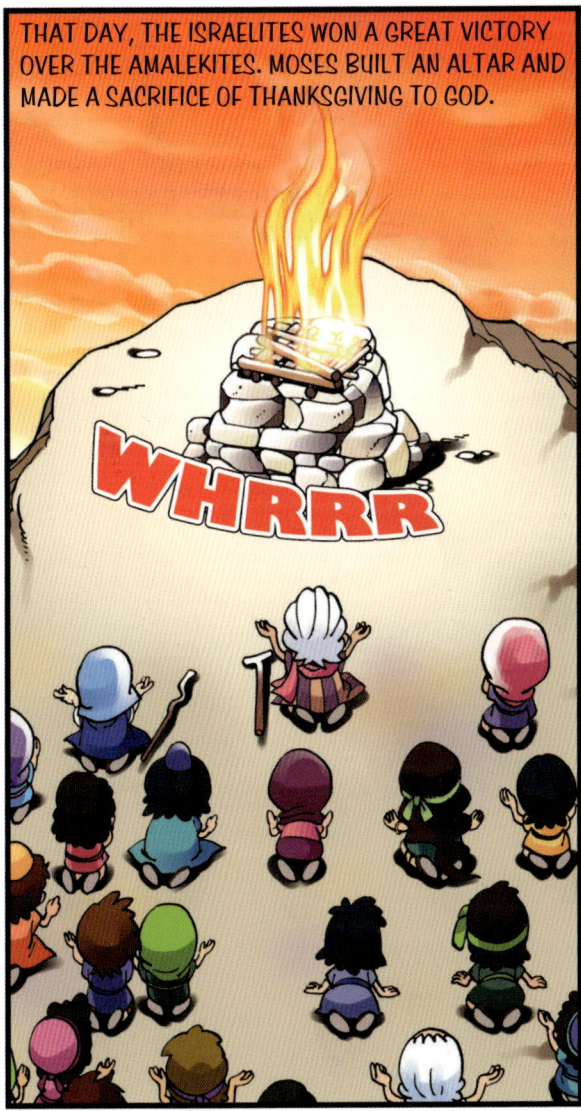

THAT DAY, THE ISRAELITES WON A GREAT VICTORY OVER THE AMALEKITES. MOSES BUILT AN ALTAR AND MADE A SACRIFICE OF THANKSGIVING TO GOD.

WHRRR

When we were fighting against the Amalekites...

I brought down a guy who was twice my size.

Oh, please. His finger was barely bigger than yours.

No, he was bigger than me by at least an arm's length.

This battle turned into a blessing. These slaves who only knew how to obey their masters rose up and defended themselves.

Yes... Now, I wish they would just stop complaining to God.

MOUNT SINAI

We will set up camp here! Arrange your tents according to your tribe! We will be here awhile, so set up well!

Moses! Your family is here! Your wife and children...

Oh! Zipporah and the kids are here...?

Surprise

Children!

Father!

Moses!

Hmm...

I have heard of all the things that you and your God have done. It is truly amazing!

Oh, Jethro, I didn't do anything. It was all the work of the Lord.

Your Lord God is surely greater than all other gods. He is the greatest in the world.

Moses, please come! We need your judgment.

?

Please rest. We just arrived here, so there are many things to do.

I will go with you. I wish to see what you do.

grumble   grumble

Hmm... This isn't good.

Moses, you alone sit as judge and have all of the people stand around you every day. Don't you get tired?

It cannot be helped. When a problem arises, the people come to me for a solution.

Well, I have a good solution for you.

You do?

Select those men who fear God, who are men of truth, and those who hate dishonest gain. Let them judge the people at all times. Let them judge every minor dispute, and only bring to you the major disputes that they cannot handle.

Thank you, my father-in-law. I will start this right away!

Ho, ho... You are always in a rush...

Oh, good idea!

Why didn't I think of that?

7. The Ten Commandments and the Golden Calf
(Exodus 19:1 – Exodus 34:35)

Hear the word of the Lord.

Everything in this world belongs to the Lord. He chose us to raise up holy sacrifices to Him.

Then are we going to be the greatest people in the world?

No, that means we are the most beloved people.

Same thing.

This means that we must keep the laws of the Lord.

Just give us His commandments! We will keep them all!

Very well. Wash your garments and keep yourselves clean for 3 days. Then, God will appear before us and speak directly to us.

Oh! I wonder what He looks like. Do you think He is scary?

I don't know, but I'm sure He's glorious and beautiful!

How will we know it's Him?

We'll know Him when we see Him.

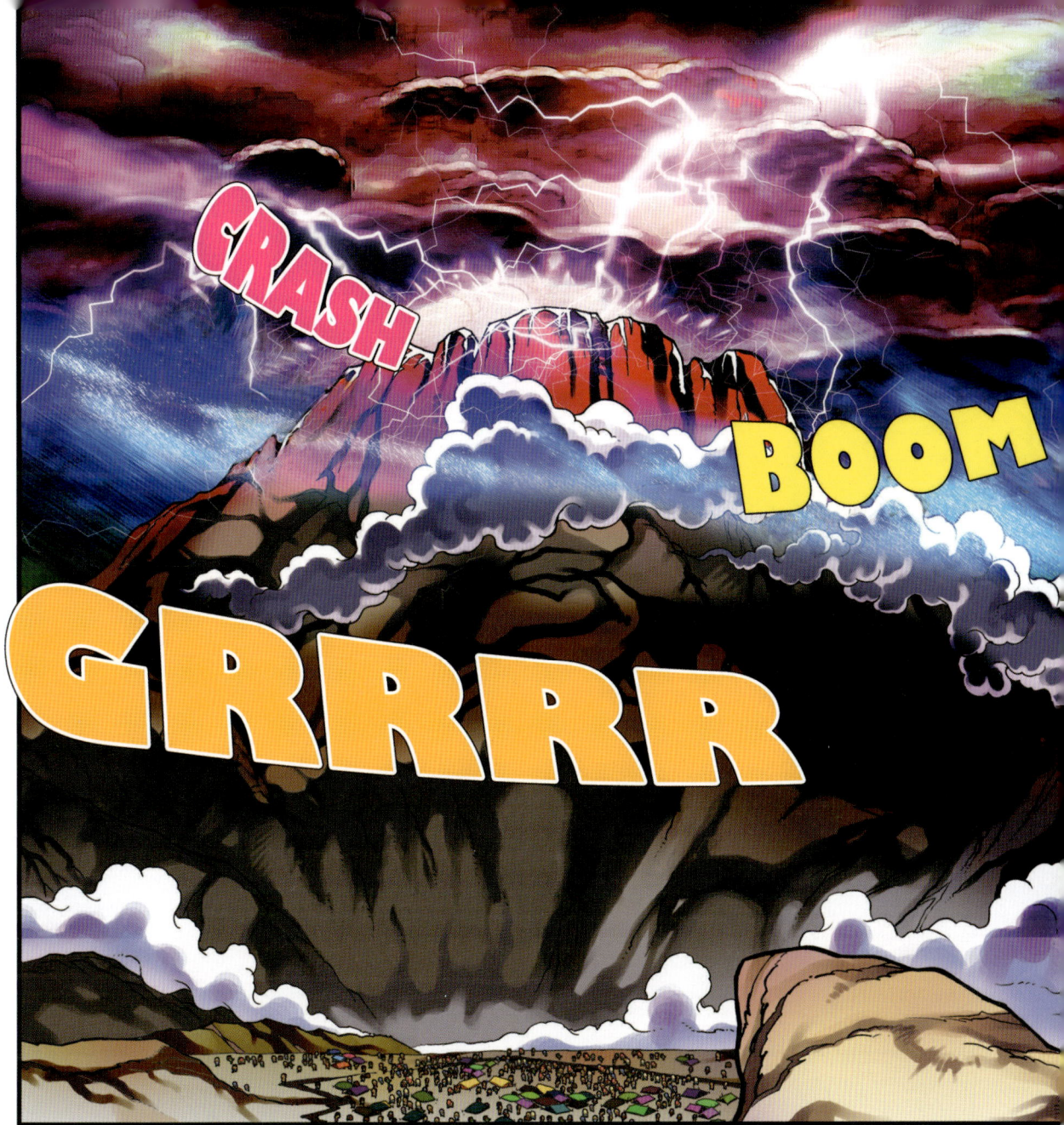

Hear the commandments of the Lord that He has just delivered!

1. I am the Lord your God. You shall have no other gods before Me.
2. You shall not make for yourself an idol, or any likeness of what is in heaven above or on the earth beneath or in the water under the earth. You shall not worship them or serve them.
3. You shall not take the name of the Lord your God in vain.
4. Remember the Sabbath day to keep it holy.
5. Honor your father and your mother.
6. You shall not murder.
7. You shall not commit adultery.
8. You shall not steal.
9. You shall not bear false witness against your neighbor.
10. You shall not covet your neighbor's house.

bustle

chatter

We will keep all of these commandments!

I shall go up the mountain to speak with the Lord further!

Joshua, come with me.

OK.

106

Sigh... I thought we were going to die.

God doesn't look like anything I've ever seen.

He is fearful, dignified, and holy... I don't know how else to describe Him.

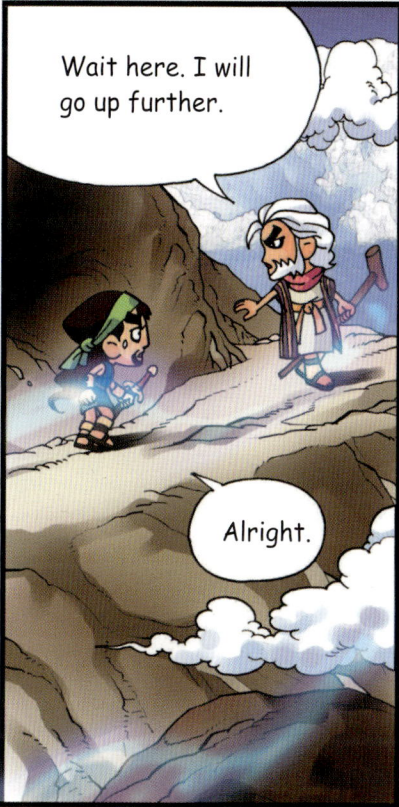

Wait here. I will go up further.

Alright.

slap slap

Lord God...

S-S-S-s

gasp

What is it?

Aaron! Do you intend for us to die in the wilderness? Make a god for us, or leave us to die here without a leader!

Very well. I will make you a god.

Aaron!

Yes! Gather all of the gold!

Bring all of the jewels!

Let's make a god in the shape of a golden calf to lead us!

What do we do...?

thump

I couldn't stand up against them.

Moses! Arise and go down the mountain at once! My people have already forgotten My commandments and they have created a golden calf to worship!

shock

What a stubborn people! I will destroy them all, and I will make of you a great nation!

O Lord! Please don't! If You destroy Your people, the foreign nations will say that You brought them out of Egypt with an evil intent, in order to kill them in the mountains.

Please turn Your anger away from them!

plop

Very well. I will withhold My anger for your sake. I shall write My commandments on two stone tablets for you. Bring them down and teach the commandments to the people.

Thank You.

S-S-S-S

....

....

shock

Moses!

Huh? That's Moses!

What? Who?

Moses is alive?

What use are these commandments for a disobedient people such as this...?!

Wait, Moses! What are you doing?

Yah!

smash

I'm sorry... I couldn't be as strong as you. I had to give them what they asked for.

Destroy that thing at once and grind it into powder. The people of Israel will drink it mixed in water! Then you will repent your sins!

112

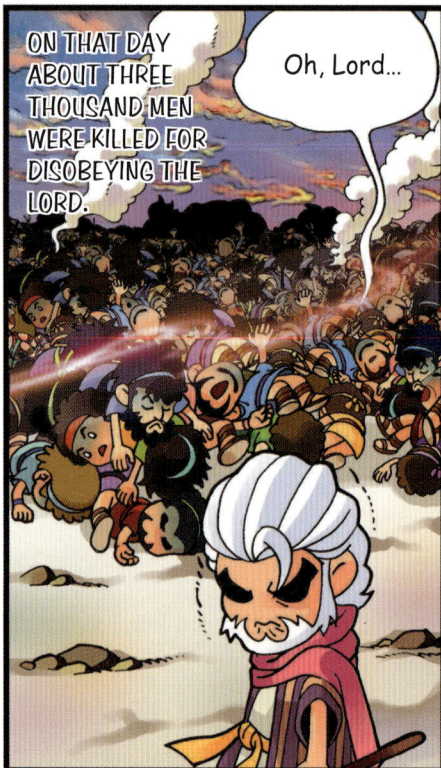

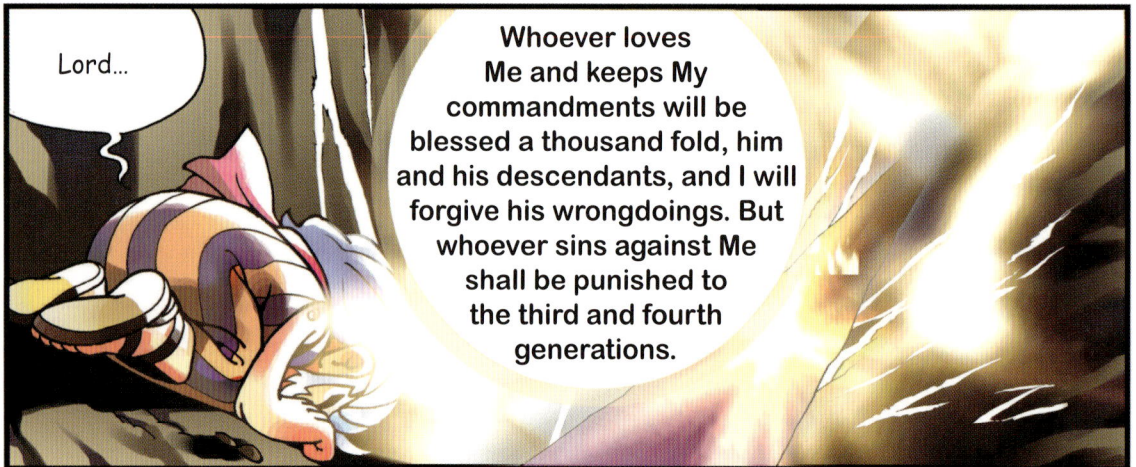

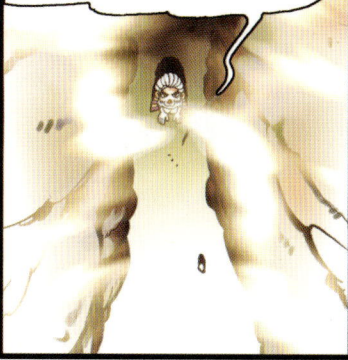

Lord... Forgive the sins of my people, and please accept them as Your people.

I am going to drive out the Amorites, the Canaanites, the Hittites, the Perizzites, the Hivites, and the Jebusites before you.

When you enter the land that I give you, be sure to keep all of My commandments.

Moses is coming!

He received God's commandments again!

Oh, my eyes!

What's going on?

SHINE

Moses' face is shining as if he was God!

Slide

The Lord has instructed us to build a tabernacle. Bring all that you are willing to contribute to the Lord. Let us combine our strength.

But if you do not wish to contribute, please do not bring anything.

I will bring my jewels.

I have animal skins...

I have lots of colored thread.

I have spices and fragrances!

I have brassware and oil...

And I have the skills to work on the construction!

But what's a tabernacle?

It's a place where God can meet with us.

SOON, THE CONSTRUCTION OF THE TABERNACLE WAS COMPLETE.

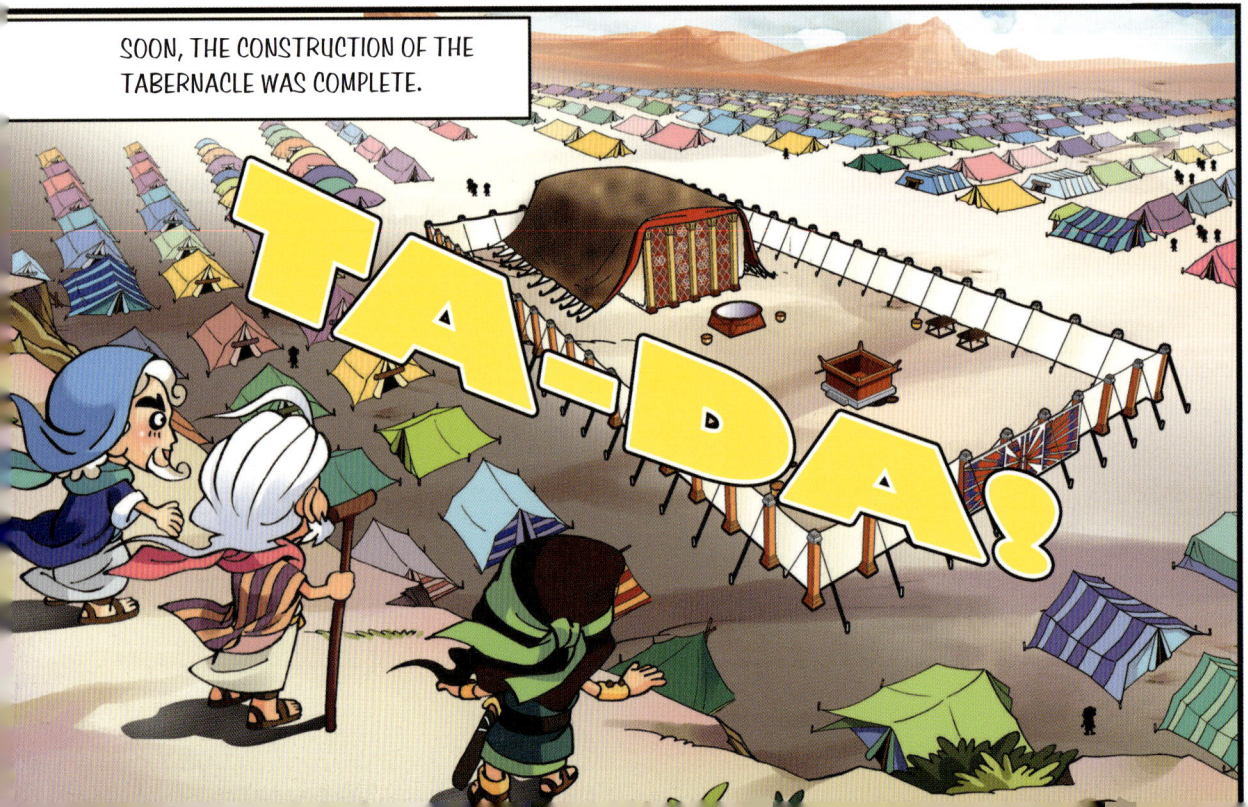

TA-DA!

INTRODUCTION TO THE TABERNACLE

THE TABERNACLE CONSISTS OF A COURTYARD AND THE SANCTUARY. THE SANCTUARY IS DIVIDED INTO THE SANCTUARY AND THE INNER SANCTUARY.

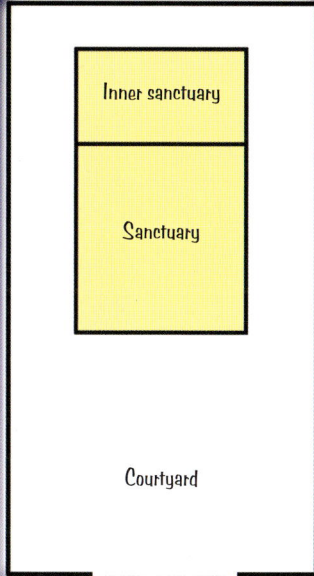

Inner sanctuary

Sanctuary

Courtyard

WHEN YOU ENTER THE EASTERN GATE, THERE IS AN ALTAR FOR MAKING SACRIFICES.

THE PRIESTS WASH THEIR HANDS IN THE LAVER*.

*LAVER: container of water

WHEN YOU ENTER THE SANCTUARY...

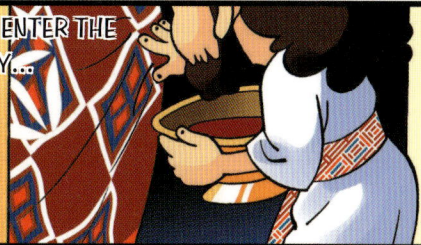

THERE IS A TABLE WITH BREAD ON THE RIGHT SIDE, A GOLDEN CANDLESTICK ON THE LEFT SIDE, AND A GOLDEN INCENSE ALTAR IN THE MIDDLE. IN THE INNER SANCTUARY, THERE ARE THE TABLETS OF THE TEN COMMANDMENTS, AND THE HIGH PRIEST ENTERS HERE TO MEET WITH GOD.

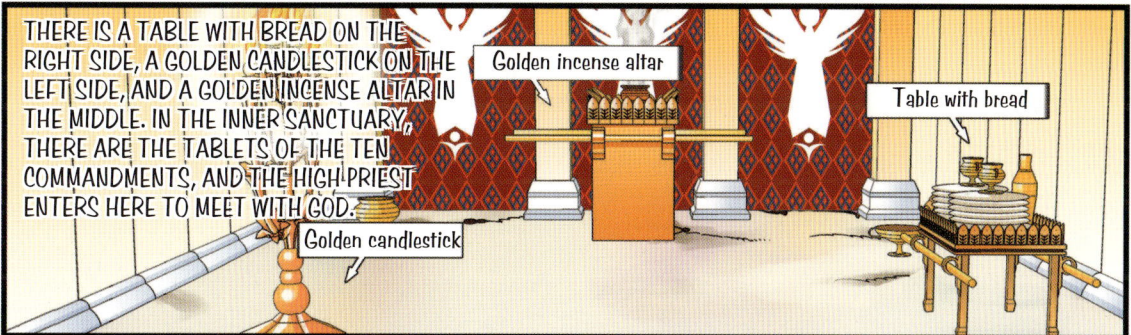

Golden incense altar

Table with bread

Golden candlestick

*To see how Aaron's rod sprouted, see pages 154-155

THE ARK OF THE COVENANT CONTAINED THE STONE TABLETS OF THE TEN COMMANDMENTS, A JAR OF MANNA, AND AARON'S ROD*, WHICH HAD SPROUTED.

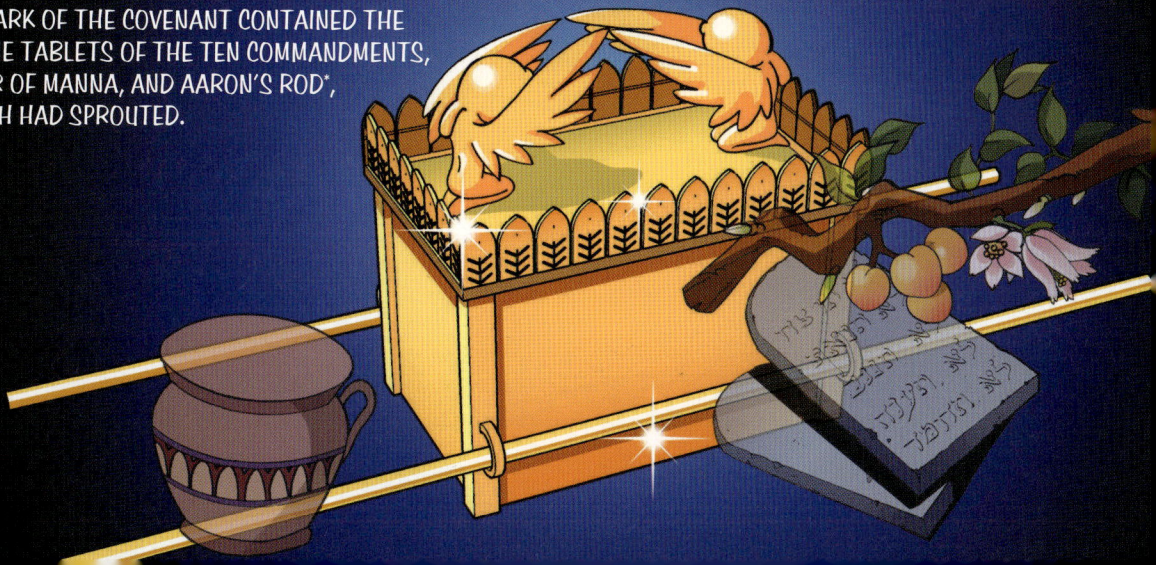

From now on, the people of Levi shall be dedicated to the Lord, and they will be in charge of all of the tasks of the tabernacle. The position of the high priest shall go to Aaron and his family.

What do you think?

sparkle

sparkle

Wonderful, Aaron!

.....

Bringing in the ark will complete the construction of the tabernacle...

TWO YEARS AFTER LEAVING EGYPT ON THE 1ST DAY OF THE 1ST MONTH, THE TABERNACLE WAS ERECTED.

SHINE

God is coming here! Oh... He is really here!

AFTER BUILDING THE TABERNACLE, THE PEOPLE OF ISRAEL LEARNED THE LORD'S LAWS AND REGULATIONS.

The Sabbath is the seventh day on which God rested after creating the world. So on that day, we must not work.

The Passover is the day when God delivered us from slavery.

The festivals are the Feast of Unleavened Bread, the Feast of Booths, and the Feast of Tabernacles.

There are many kinds of offerings— burnt offerings, grain offerings, peace offerings, sin offerings, guilt offerings... So hard to remember!

The murderer must be killed, and the thief...

God hates idols above everything else. Do not worship them!

MOSES TAUGHT THE PEOPLE EVERYTHING THAT HE HAD LEARNED FROM GOD.

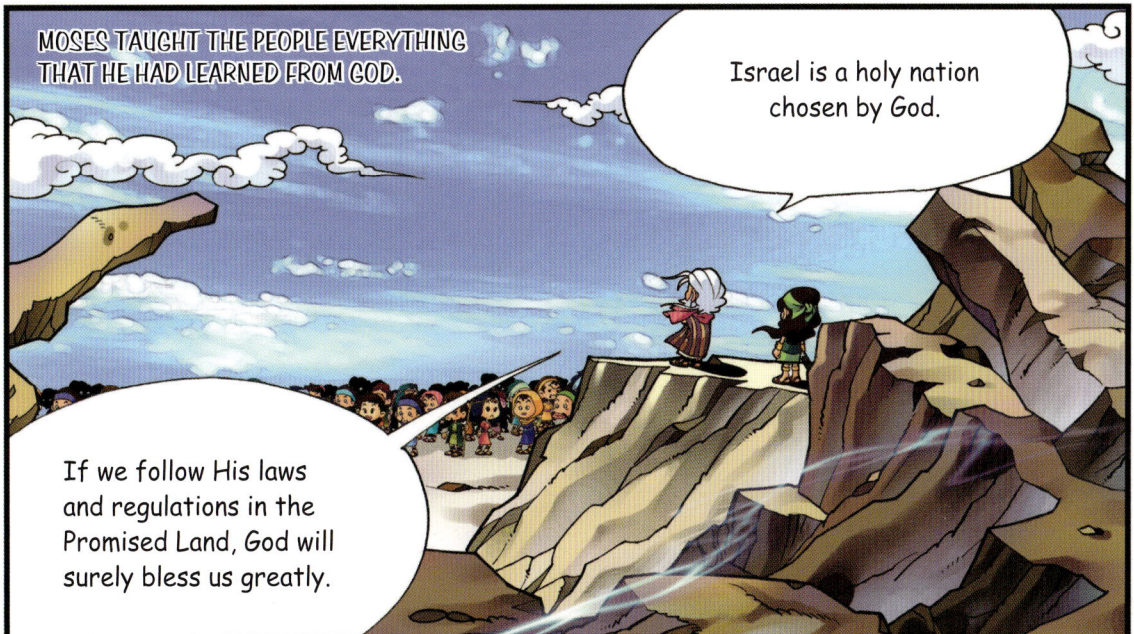

Israel is a holy nation chosen by God.

If we follow His laws and regulations in the Promised Land, God will surely bless us greatly.

Is that the special fire?

yikes

Well, fire is all the same. Is some holier than others?

Hee hee hee

Heh-heh, I have it, too.

Are you going to burn incense?

Yes.

Go with pure hearts.

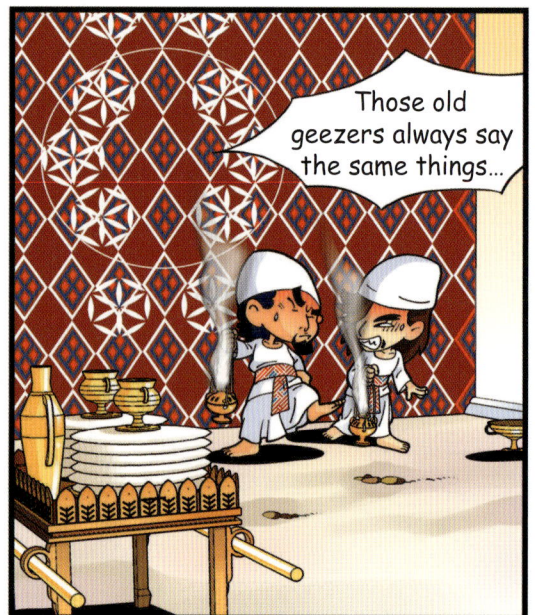

Those old geezers always say the same things...

Nadab!

Abihu!

My sons...!
How did this happen...?

SSSS

They didn't use the fire that God had ordered.

God's law is absolute. I guess there is nothing more to say about the rules of the tabernacle.

THEN THE LORD SPOKE TO MOSES, "DEPART, GO
UP FROM HERE, YOU AND THE PEOPLE WHOM
YOU HAVE BROUGHT UP FROM THE LAND OF
EGYPT, TO THE LAND OF WHICH I SWORE TO
ABRAHAM, ISAAC, AND JACOB, SAYING, 'TO
YOUR DESCENDANTS I WILL GIVE IT.' "
EXODUS 33:1 (NASB®)

Chapter 3

Toward the Promised Land

HAZEROTH

ALL PEOPLE ARE AMAZED BY THE LORD'S POWER, AND THEY ARE AFRAID.

pause

The ark has stopped!

O, Lord! Please return to Your people!

How long are we going to march like this...?

I'm sick and tired of walking in the wilderness.

And all there is to eat is manna.

crackle
crackle

O Lord...!

thump

This weight is too heavy for me to bear.

They are not my children. But why must I lead them by myself to the land that You promised to our ancestors...?

All they do is cry for meat and complain. I cannot bear to look at these people any longer. You promised to bless me, yet You make me suffer like this. It would be better if You killed me now.

Fire!

gasp

O Lord...

The fire went out!

The Lord has heard your complaints. He will give you meat, just as you desire.

Really?

We will eat meat?

Yay!

Thank You, Lord!

You wouldn't be so happy if you knew what God intended for you...

Look at Moses. Does he think that God only speaks through him? I'm a prophetess, too. And you're the high priest.

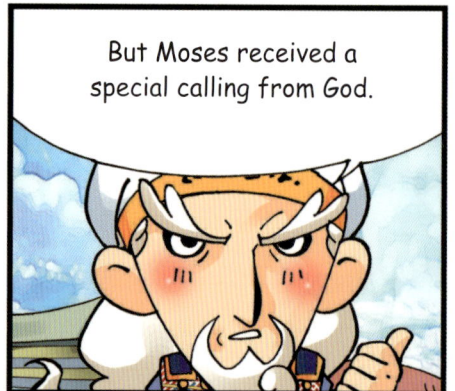

But Moses received a special calling from God.

Moses, I need to talk to you.

We also have heard the word of God. But these days, it seems that you decide everything on your own and ignore us.

Well, sister...

hmmm

AT THAT MOMENT, THE LORD CALLED THE THREE OF THEM.

136

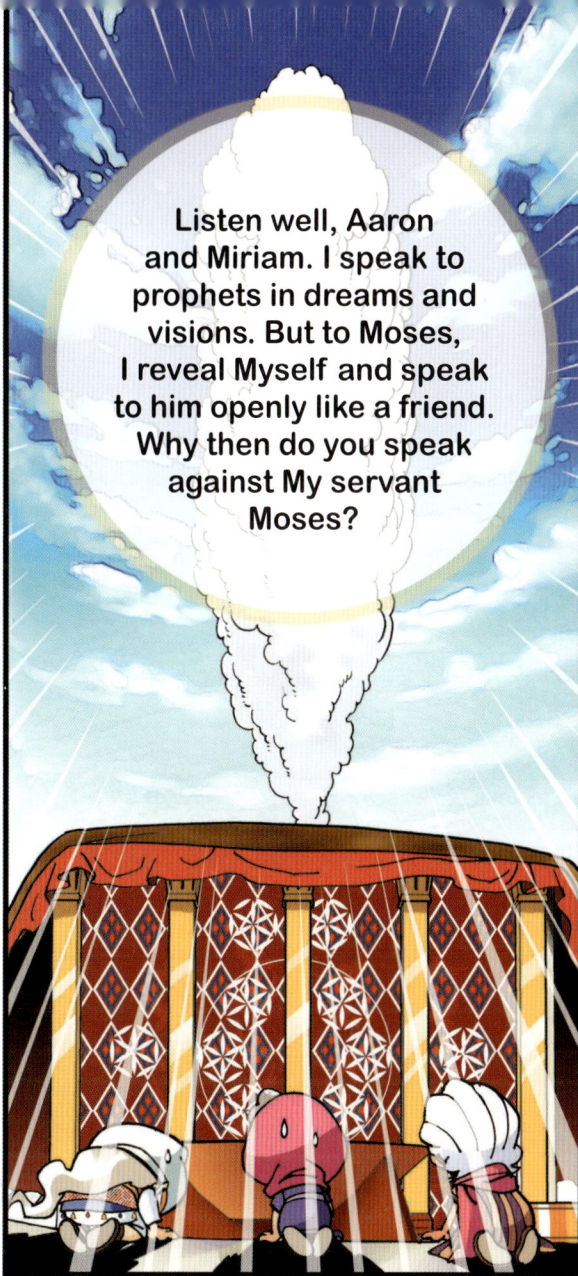

Listen well, Aaron and Miriam. I speak to prophets in dreams and visions. But to Moses, I reveal Myself and speak to him openly like a friend. Why then do you speak against My servant Moses?

Oh! My hands!

I have leprosy!

Aah!

I'm sorry, Lord! Please forgive me!

We were foolish! Moses! Please forgive us, and pray to God on our behalf!

Lord...! Please heal my sister!

MIRIAM HAD LEPROSY FOR 7 DAYS AS PUNISHMENT FOR SPEAKING AGAINST MOSES. SHE HAD TO REPENT OUTSIDE OF THE CAMP.

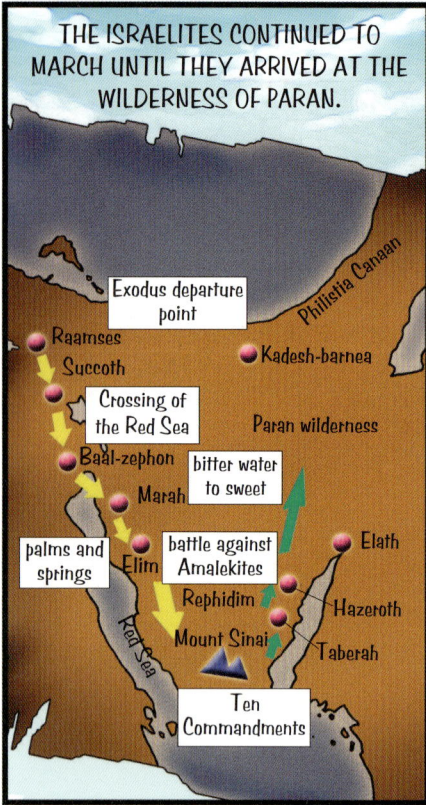

THE ISRAELITES CONTINUED TO MARCH UNTIL THEY ARRIVED AT THE WILDERNESS OF PARAN.

Ba-room!

I hear the silver trumpets. What's going on?

It was just one blast. So they're calling the leaders of the tribes.

You will go now and scout out the land of Canaan, where we will live.

Yay!

Oh! We're finally here! We waited for this day for so long!

Yeah!

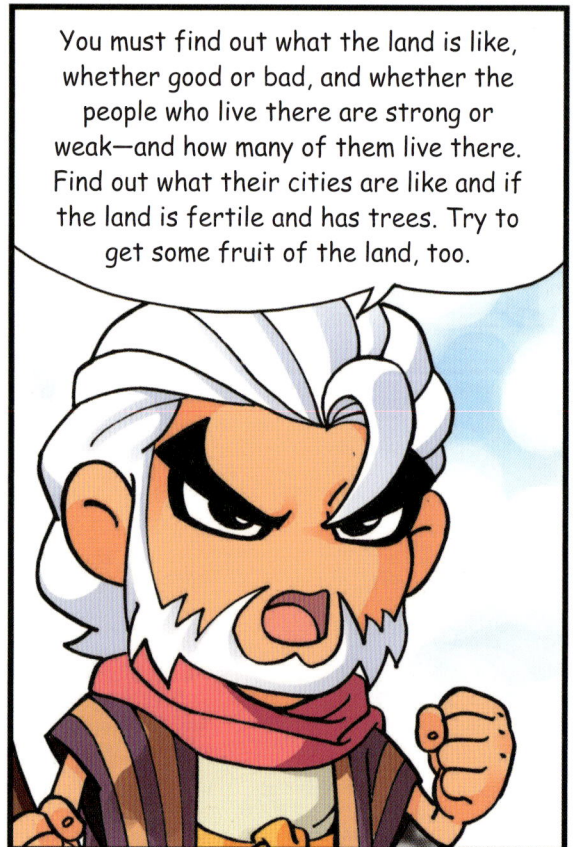

You must find out what the land is like, whether good or bad, and whether the people who live there are strong or weak—and how many of them live there. Find out what their cities are like and if the land is fertile and has trees. Try to get some fruit of the land, too.

Let's head out all over the land.

We will head north.

We will go east.

Then we will go west.

Let's go, too!

Joshua. My heart is already pounding.

Don't worry, Caleb. The land that God gives us will surely be amazing.

Look!

Look at the size of these grapes! Let's take some on the way back.

If the fruit grows this big, that means the land is very rich and fertile, right?

I knew it. Our God is really amazing!

40 DAYS LATER, ISRAELITE CAMP
Look! The scouts have returned!
bustle
bustle
Look at those grapes! They're huge!
Look, Moses! Isn't it amazing?
The land is rich and fertile. There is much to eat, just like these grapes.
That's true, but the land is occupied by the descendants of Anak, who are very tall and strong.
The cities are fortified and very large!
Who cares if the land is plentiful. It's full of very strong people...
We will be killed before we can even enter.
Our children will be taken as slaves.
It would be better to die in Egypt or the wilderness than to be killed by their swords.

Do you plan on sacrificing us to the Anak people?

We'll stone you before that happens!

....

Kill Joshua and Caleb!

Let's pick new leaders and go back to Egypt!

Look! The tabernacle...

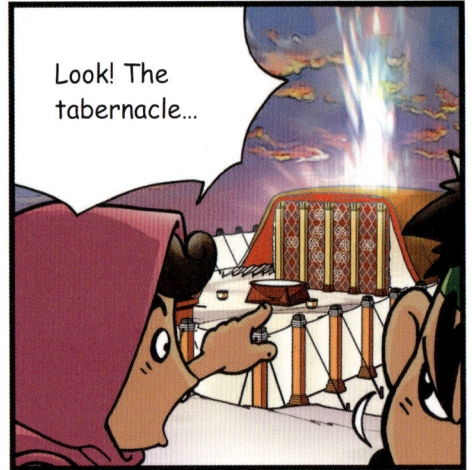

God is here!

chatter

chatter

What do we do? Do you think He saw us fighting?

How long will these people not obey Me?!

Just as an eagle protects its young, I have saved you from Egypt. I have performed countless miracles, yet they still rebel against Me and reject the land that I have promised to them.
I will kill them all and establish a new nation through you.
Lord...
Please don't do such a thing! The Egyptians and the foreign nations all know of the great things that You have done for our people. But if You give up on us now, they will say that You could not fulfill Your promise.
Please take away Your anger and forgive us in Your love.
I will forgive them for your sake.
But they have rebelled against Me too many times. As I live, not a single one among them will enter the Promised Land!

cock-a-doodle-doo

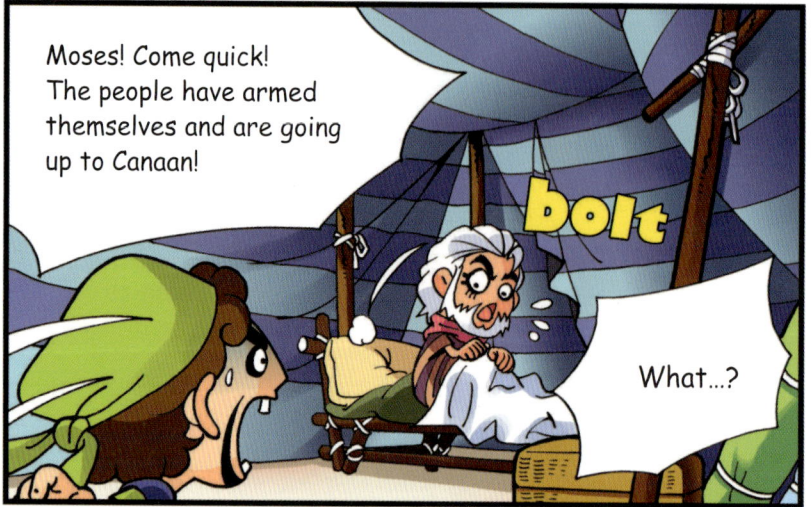

Moses! Come quick! The people have armed themselves and are going up to Canaan!

bolt

What...?

STOP!

Don't try to stop us! We repented all night!

Now, we will go up and take the land!

Will you disobey Him again?

The Lord has commanded us to return to the wilderness!

146

THE PEOPLE DISOBEYED THE LORD AND WENT UP TO CANAAN. THEY LOST THE FIGHT, AND ONLY A FEW RETURNED ALIVE.

THE DISOBEDIENT PEOPLE ALSO REFUSED TO BELIEVE MOSES.

Can we really believe everything that Moses says?

Even if he lies and says they are God's words, we would never know.

Plus, we Levites are in charge of the Lord's tabernacle. We don't need to follow Moses!

Why should the position of the high priest only go to the line of Aaron?

Let's show the people tomorrow who should become the new leader!

Good idea.

Moses!

We won't follow you anymore.

We were happy in Egypt, but you dragged us out. You promised us a land of milk and honey, but now you are going to make us perish in the wilderness!

Ha! You must be scared, to leave this until tomorrow...

We will give you time to say your final good-byes. Tomorrow morning, you will be driven out.

Well? Why don't you try calling the Lord?!

Korah...!

Do you know what you are doing?

150

Of course, I do! By this time tomorrow, I will be the new leader of this people.

Now, get rid of him!

All of those who wish to live, back away from Korah, Dathan, and Abiram! The Lord is greatly angered!

You wish...

I'm tired of your warnings.

What do we do?

I will believe Moses.

I will believe Korah.

I... don't know... Oh, man...

chatter

chatter

Hurry!
Yikes! I'm going to Moses!
I don't know anymore! I'm just going to run!
CRACK
Aah!
ZOOM
Someone, save me!
Oh! Look!
WHRRRR
Oh, no!
Argh!

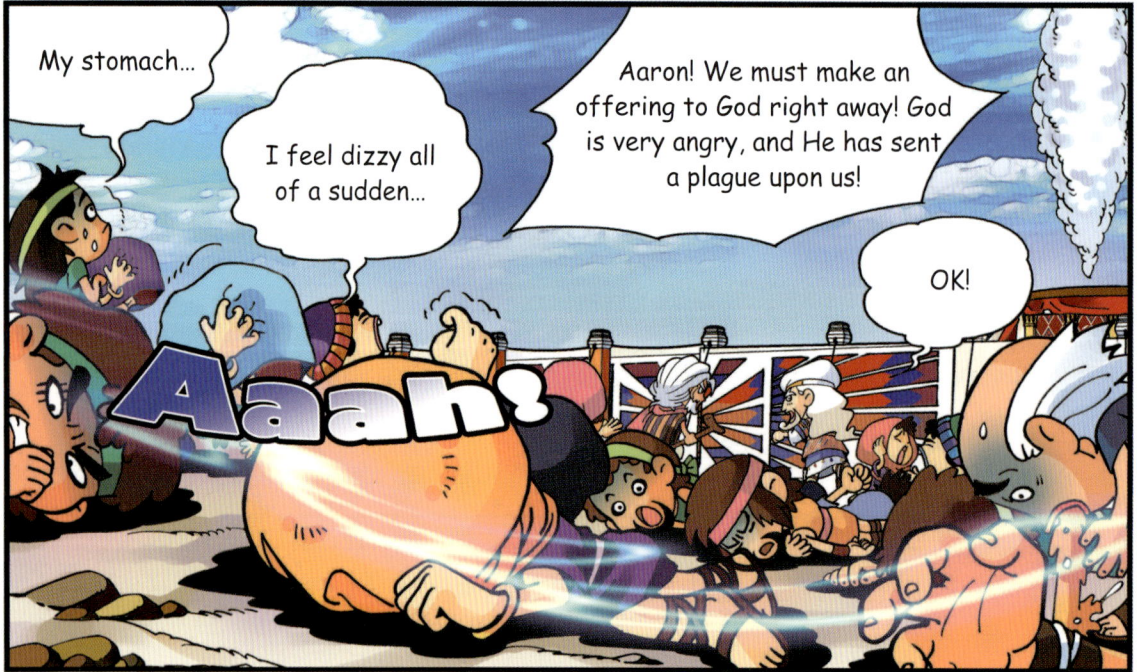

My stomach...

I feel dizzy all of a sudden...

Aaah!

Aaron! We must make an offering to God right away! God is very angry, and He has sent a plague upon us!

OK!

THAT DAY, MORE THAN 14,700 PEOPLE DIED FROM THE PLAGUE. BUT EVEN AFTER THAT, THE PEOPLE OF ISRAEL CONTINUED TO SPEAK AGAINST MOSES AND AARON.

Couldn't they have made the offering more quickly?

They delayed the offering on purpose to make us suffer more...

AND SO, GOD SHOWED THEM ONE MORE SIGN.

Leaders of the twelve tribes, please take your rods and write your names on them.

We will place the rods in the ark of the covenant. The rod of the one chosen by God will sprout.

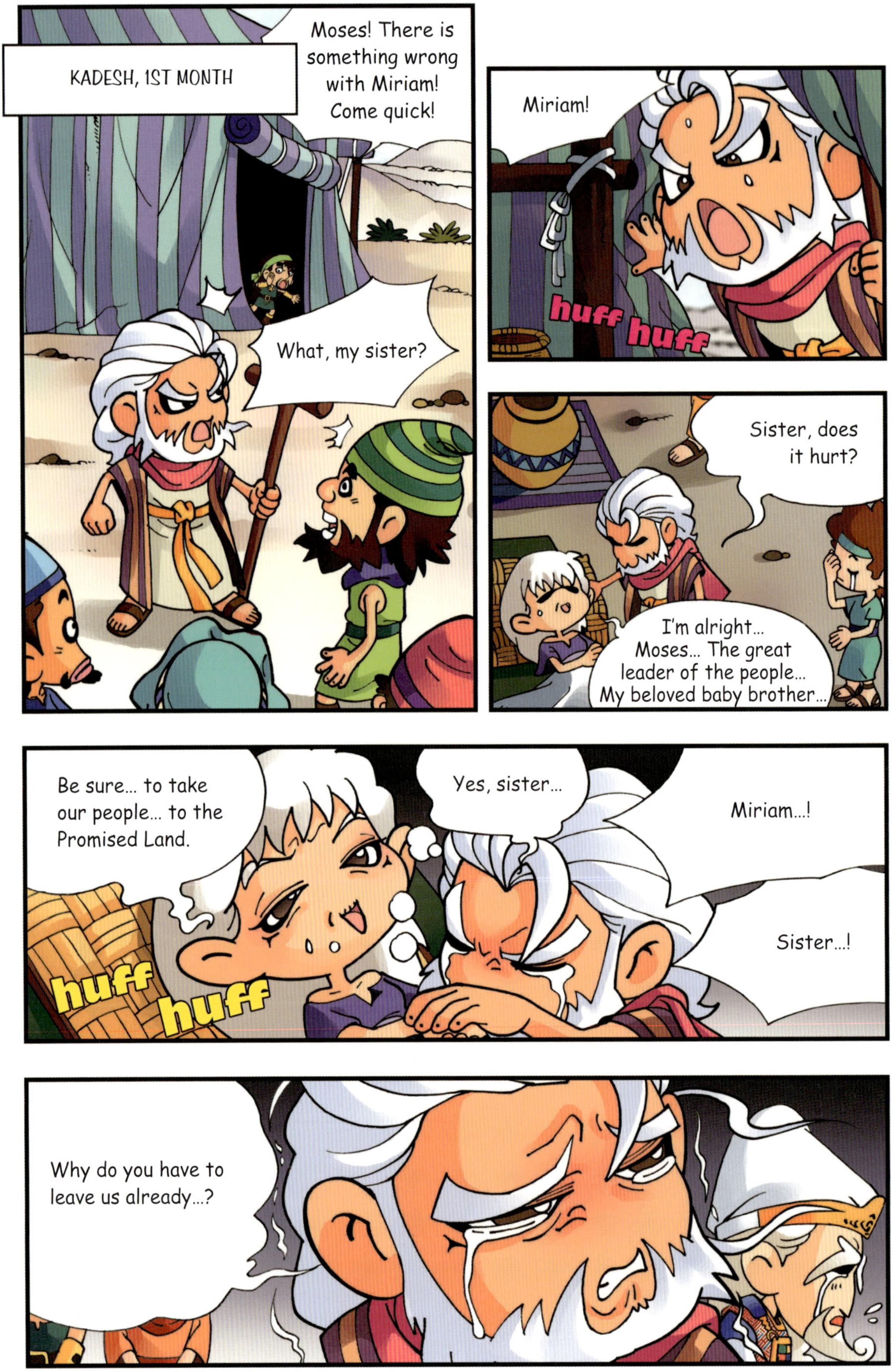
KADESH, 1ST MONTH
Moses! There is something wrong with Miriam! Come quick!
What, my sister?
Miriam!
huff huff
Sister, does it hurt?
I'm alright... Moses... The great leader of the people... My beloved baby brother...
Be sure... to take our people... to the Promised Land.
Yes, sister...
Miriam...!
Sister...!
huff huff
Why do you have to leave us already...?

159

God told me to say
something else.

Then why...

I was so angry that
I couldn't help
myself.

sigh

I am tired of all this.

Moses... Even
you grow tired...

Moses! You did not show My glory to the people.

Forgive me, Lord. I have sinned against You.

Therefore, you will not enter the Promised Land!

Lord...! Are You serious...?

tremble

Brother, bring Eleazar here.

My son Eleazar? Why...?

We must pass onto him the role of the high priest.

That means...

huff

huff

Yes, it has been a long life...

AARON, THE HIGH PRIEST AND THE BROTHER OF MOSES, DIED ON MOUNT HOR AT THE AGE OF 123.

MOSES CLOTHED AARON'S SON ELEAZAR IN THE GARMENTS OF THE HIGH PRIEST AND RAISED HIM UP AS THE NEW HIGH PRIEST.

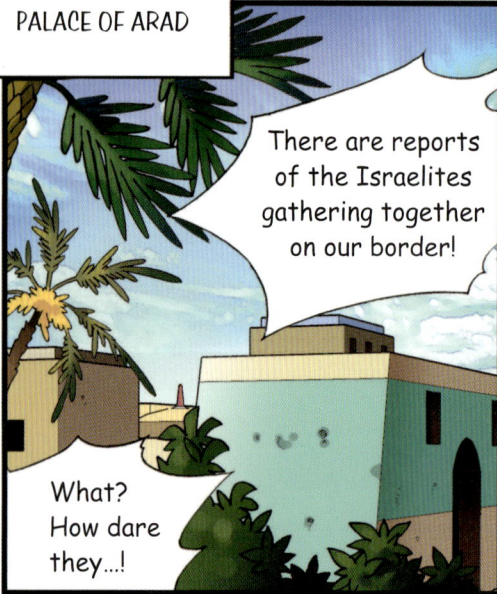

PALACE OF ARAD

There are reports of the Israelites gathering together on our border!

What? How dare they...!

Go and destroy them all! Show them the terror of Arad!

dong dong

Some of our soldiers have been captured by the Arad army!

Lord God! If You will help us, we will completely destroy Arad!

GOD HEARD THE PLEDGE OF THE PEOPLE, AND HE ALLOWED THEM TO DEFEAT ARAD.

Hallelujah!

Praise God!

We won!

BUT WHEN THEY RETURNED TO THE WILDERNESS IN ORDER TO GO AROUND EDOM, THE PEOPLE COMPLAINED AGAIN.

We defeated Arad. Why can't we live there?

How humiliating...

We have to go back the way we came!

splat

I'm sick of kneading manna!

I miss Egypt.

We're out of manna recipes.

It wouldn't be so bad if we had at least a lot of water to drink...

Oh! A snake!

gasp

They're poisonous!

The poisonous snakes are everywhere!

Moses! Save us!

HISS

We were wrong! We won't complain about you or God anymore! Please, forgive us!

If you were bitten, look at this bronze serpent and you will live!

Bronze serpent?

Oh, amazing! I'm all better.

I was healed just by looking at it!

Thank You, God! I won't doubt You ever again.

HESHBON, PALACE OF SIHON

Hmm... They wish to go through our land...?

WITH EVERY BATTLE, THE ISRAELITES GREW STRONGER. NEWS OF THEM SPREAD THROUGHOUT THE LAND.

We are the strongest! Hurrah for the army of God!

GOD WAS WITH THE ISRAELITES, AND THEY FEARED NOTHING.

The God of Israel is a great and powerful God!

He even parted the sea...

...and He sends food down from heaven.

As long as God is with us, Israel is the strongest!

That is Og, the King of Bashan. He has heard of us and has come out already.

Hmm... They are large in number.

Is the King of Bashan a giant?

They say that his bed is made of steel.

And it's more than 13 feet long.

The walls of Bashan are high and strong...

Can we really win?

Do not be afraid! God fights with us! He has already given Bashan over to us!

Let's go!

RAAH

WITH GOD'S HELP, THE ISRAELITES COMPLETELY CONQUERED THE 60 CASTLE TOWNS OF BASHAN. THEN, THEY CAME TO SETTLE IN THE PLAINS OF MOAB IN PREPARATION FOR CROSSING THE JORDAN RIVER.

Thank goodness we're finally done with that endless wilderness.

We're finally here. Once we cross the Jordan River, it's the Promised Land.

11. The False Prophet Balaam
(Numbers 22:1 – Numbers 25:18)

PALACE OF BALAK, KING OF MOAB

rush

Sire! The Israelites have set up camp on the plains of Moab! There are many of them, just like the rumors said!

They are finally here. It is just a matter of time until they attack.

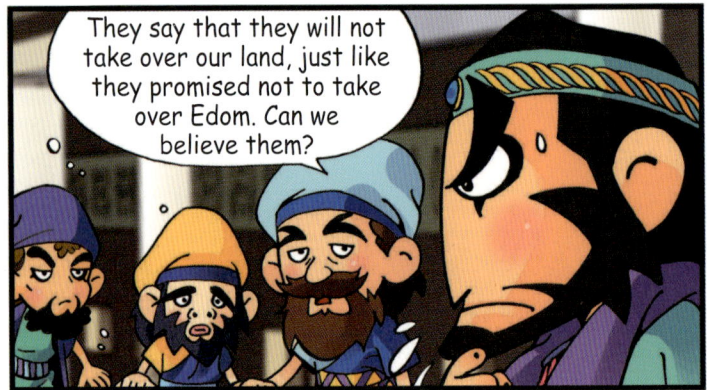

They say that they will not take over our land, just like they promised not to take over Edom. Can we believe them?

Have you ever seen a cow pass through a green pasture without tasting the grass?

No, it will eat the grass, and trample the ground... what should we do?

Sire! Do you remember the Prophet Balaam, the son of Beor?

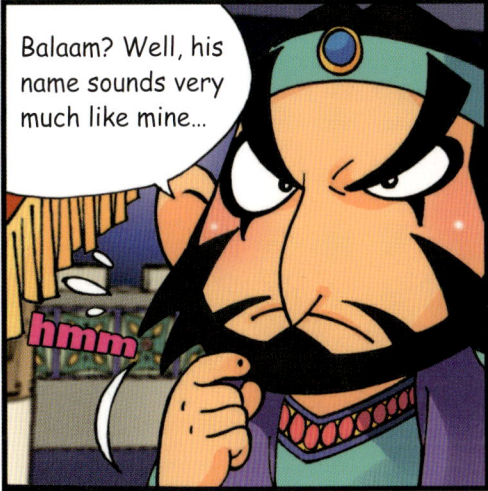

Balaam? Well, his name sounds very much like mine...

hmm

They say that his words are powerful.

Whomever he blesses is greatly blessed, and whomever he curses is greatly cursed.

Now that you mention it, I think I have heard of him.

He is from your hometown.

How can you not remember someone from your own hometown?

Very well! Go with the elders of Midian and bring Balaam to me.

I already know why you are here.

Balak is afraid of Israel, right?

Oh~ You really are an amazing prophet!

We can completely trust you!

Everyone has heard the news about Israel...

Then, what shall I do for you?

Very well. Spend the night here, and I will let you know what the Lord says.

Come with us and curse Israel! Then, we will be able to drive them out of our land!

BALAK, KING OF MOAB

175

176

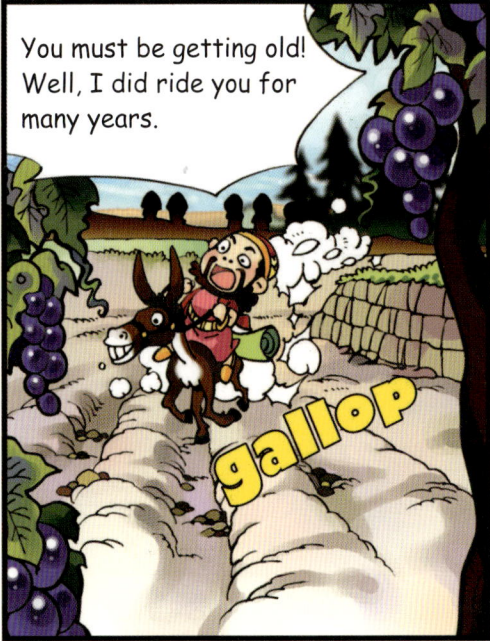

You must be getting old! Well, I did ride you for many years.

gallop

Hee-haw! Hee-haw!

Ouch!

screech

splat

Why did you stop like that? What's wrong with you?

Why are you hitting me? What did I do wrong?

Gasp! My donkey is... talking?!

Stay calm...!

If I had a sword, you would already be dead!

Now, I'm talking to a donkey! Oh, dear...

What?!

Let's set the record straight, Master. I have served you faithfully all of my life. Never once have I disobeyed your command.

Well, yes, you're right. *hiccup*

Oh, he can speak really well.

It's an...

shock

...angel!

Foolish man! Your donkey saw me as I blocked your path. If it had not gone off the path, I would have killed you with my sword.

I'm sorry!

I didn't see you, angel of the Lord!

If God is unhappy with me going to Moab, then I will go back right now.

No. Go with them to Moab. But you must say only the words that God says to you. Do you understand?

Yes.

?

You've come.

I've waited for so long.

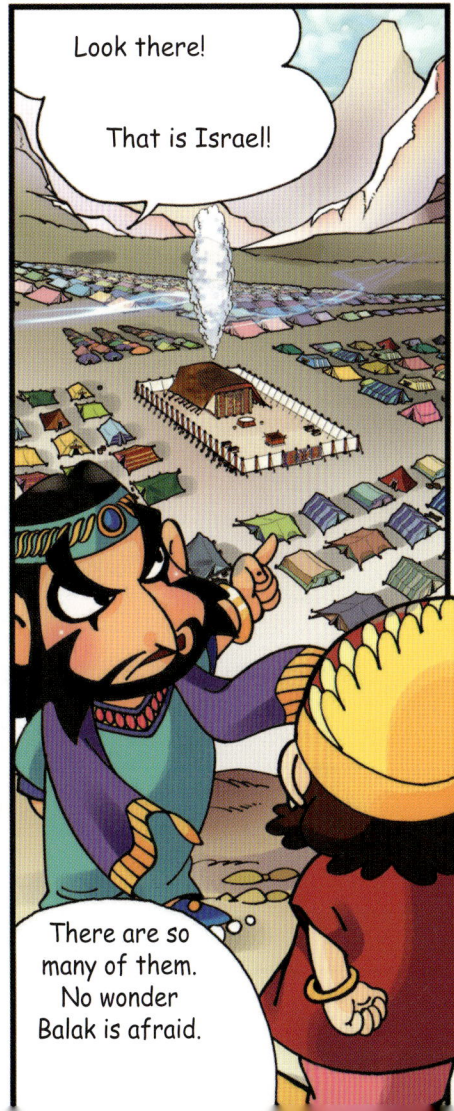

Look there!

That is Israel!

There are so many of them. No wonder Balak is afraid.

Behold, a people who live apart and who will not be counted among the nations.

snap

Hmm... That means that they are extremely self-willed!

That must be bad, right?

How shall I curse those whom God has not cursed?

Huh?

Who can count the dust of Jacob, or number the fourth part of Israel?

Let me die the death of the upright, and let my end be like his!

What is this...?!

Balaam! What is the meaning of this?! I asked you to curse them, but instead, you give them a blessing!

Did he get the blessing and the curse mixed up?!

If not, is he afraid...?

He might be a fake prophet.

I can say only what the Lord puts in my mouth.

So it's not my fault!

Maybe Balaam is afraid because there are so many of them.

You think...?

Let us go to another place from where you will see only a few of them. Then, curse them for me there!

If you wish. Build seven more altars there as you did before and prepare the sacrifices.

Now, you should be fine with cursing them from here!

Wait here next to the offering. I will go speak with the Lord.

What if he says the wrong thing again?

slap slap slap

No, it must have been a mistake.

The Lord has given His message to me.

Good. Let it be a really big curse this time...

Remember. God is not a human being, so He does not lie or have regrets. He definitely does what He says He will do.

That's why I'm asking for a curse here.

The Lord God is with Israel, and the shout of a king is among them.

All nations will praise the mighty workings of the Lord for His people.

Behold, they rise like a lioness, and as a lion lifts itself; it will not lie down until it devours the prey.

Argh... What's wrong with him? Is he really a fake after all?

Urgh... Again?!

If you can't curse them, then at least you should stop blessing them. What's going on?

GRRRR

I can say only what the Lord commands me to say, right? So it's not my fault!

He must not have any powers after all.

Rumors are just rumors.

Why don't we get someone else?

But since it was difficult to get him here, let us try one more time.

Let us go to the top of Mount Peor! Perhaps God will allow you to say a curse from that spot.

As you wish. But I can only do what the Lord commands.

They are well organized into tribes.

They are quite an organized army... They only could have done this with God's help.

*ALOE: a healing plant

His seed will be by many waters...

Stop it! I don't want any more blessings...!

Be quiet!

And His kingdom shall be highly praised!

Aaah

God is for him, and he will conquer the nations who are his enemies!

Get him! Stop him from talking!

run

O Israel, blessed is everyone who blesses you, and cursed is everyone who curses you!

We're done for...!

thump

Get him out of here, right now!

Don't be so angry, my King!

Remember, I said that I would deliver only the message of the Lord, even if you gave me a palace full of gold. I am just keeping my promise.

Believe me, this is not my will!

You have until the count of three to disappear!

One.

TWO.

GRRRR

OK! I'm going!

pause

turn

run

Three...

One last thing!

I have a prophecy about Israel's distant future!

A star shall come forth from Jacob, a scepter* shall rise from Israel, and shall crush through the forehead of Moab, and tear down all the sons of Sheth. Edom shall be a possession,

Seir, its enemies, also will be a possession, while Israel performs bravely. Alas, who can live unless God has ordained it?!

* SCEPTER: a staff carried by a king to show his authority

Kill him!

No way~!

Swoosh

Swoosh

Looks like fun!

Want a bite?

Well...

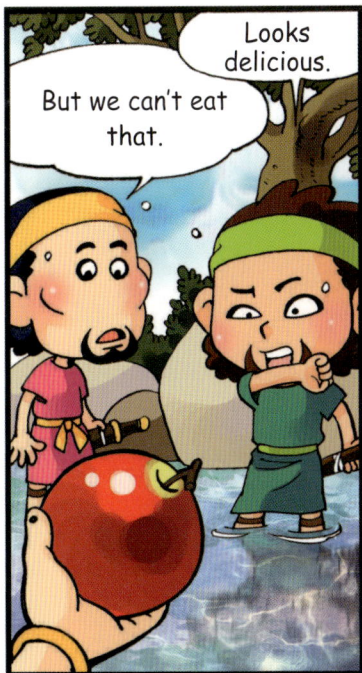

Looks delicious.

But we can't eat that.

Oh! I forgot. You only can eat manna, right?

ho ho ho

Ha-ha, you're cowards!

ha ha ha

Do you think it's poisoned?

Well, if it's just once...

No one's looking.

OK, just once!

WHILE ISRAEL REMAINED AT SHITTIM, THEY BEGAN TO WORSHIP BAAL, THE GOD OF MOAB.

The festival for Baal is so much more fun than God's festival.

The sacrifices to the Lord are too strict.

Yeah

And we can eat all of the offerings to Baal without being killed!

THE LORD WAS GREATLY ANGERED. HE SENT A TERRIBLE PLAGUE UPON THE PEOPLE AS PUNISHMENT FOR THEIR BETRAYAL.

These people...!

I've told you time and time again not to worship idols! How could you let this happen? Do you think that you can escape the wrath of God?!

All of those who bowed down to the god of Moab will be executed!

I've never seen Moses so angry.

He was mad at Kadesh, but not like this.

What do I do? My son was one of the people who bowed down...

Who will lead in carrying out the Lord's command?

I can't...

Me, either...

SWISH

I will.

Phinehas! Will you do this?

You have risen up in bravery for the honor of the Lord. Therefore, God will bless you, and your descendants always will have the priesthood!

PHINEHAS, THE GRANDSON OF AARON, AROSE. HE EXECUTED THE LEADER OF THE PEOPLE WHO HAD SINNED AGAINST THE LORD THROUGH MOAB. THEN, THE PLAGUE THAT HAD BEEN SPREADING THROUGHOUT ISRAEL STOPPED. HOWEVER, MORE THAN 24,000 PEOPLE HAD DIED ALREADY FROM IT.

Just look at us.

We're dying with the Promised Land right in front us...

We forgot that God hates idol worship more than anything else... Who knew that it would be such a tremendous sin...?

THE INCIDENT AT PEOR AWAKENED THE ISRAELITES, AND THEY DEVELOPED A STRONG FAITH THAT WOULD BRING THEM INTO THE PROMISED LAND.

Hmm. I think we can enter Canaan now.

12. The Death of Moses and a New Leader
   (Numbers 27:12 – Numbers 27:23) (Deuteronomy 31:1 – Deuteronomy 34:12)

Lord, we are now ready to enter the land that You promised to our forefathers.

Lord, You raised up Israel from Egypt with a mighty hand and made us Your holy nation. Wherever we went, all nations were afraid of Your great and mighty name.

We complained and rebelled countless times, but You did not abandon Your people. You led us here until the very end.

O Lord, please hear my plea...

Please let me enter the Promised Land. I would like to see and touch the milk and honey flowing there just once.

Moses.

Yes, Lord.

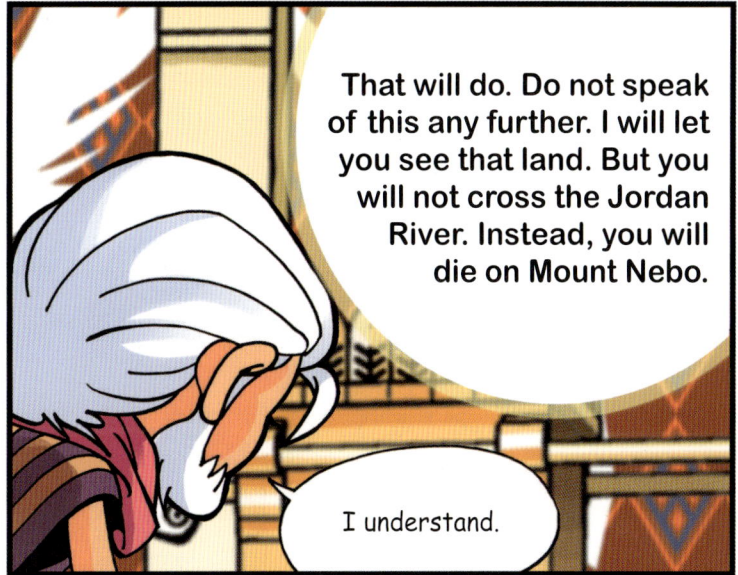

That will do. Do not speak of this any further. I will let you see that land. But you will not cross the Jordan River. Instead, you will die on Mount Nebo.

I understand.

Everyone! Listen carefully to my words, and be sure to pass them down to your children.

Love the Lord your God with all your heart, soul, and mind!

This is God's greatest commandment.

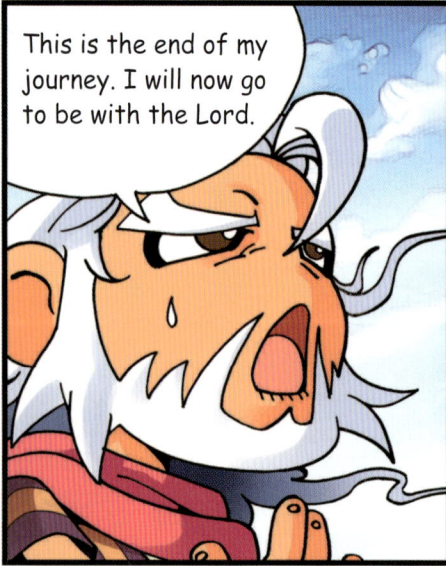

This is the end of my journey. I will now go to be with the Lord.

Then, who will we follow into the Promised Land?

Don't we need a leader?

God has already chosen the next leader.

It is Joshua!

Strengthen your heart and lead Israel well!

Yes...! I will obey the Lord.

MOSES SAID FAREWELL TO THE PEOPLE OF ISRAEL, AND HE WENT UP MOUNT NEBO AS THE LORD COMMANDED.

THERE, HE COULD SEE ALL OF THE PROMISED LAND.

huff

huff

That land...

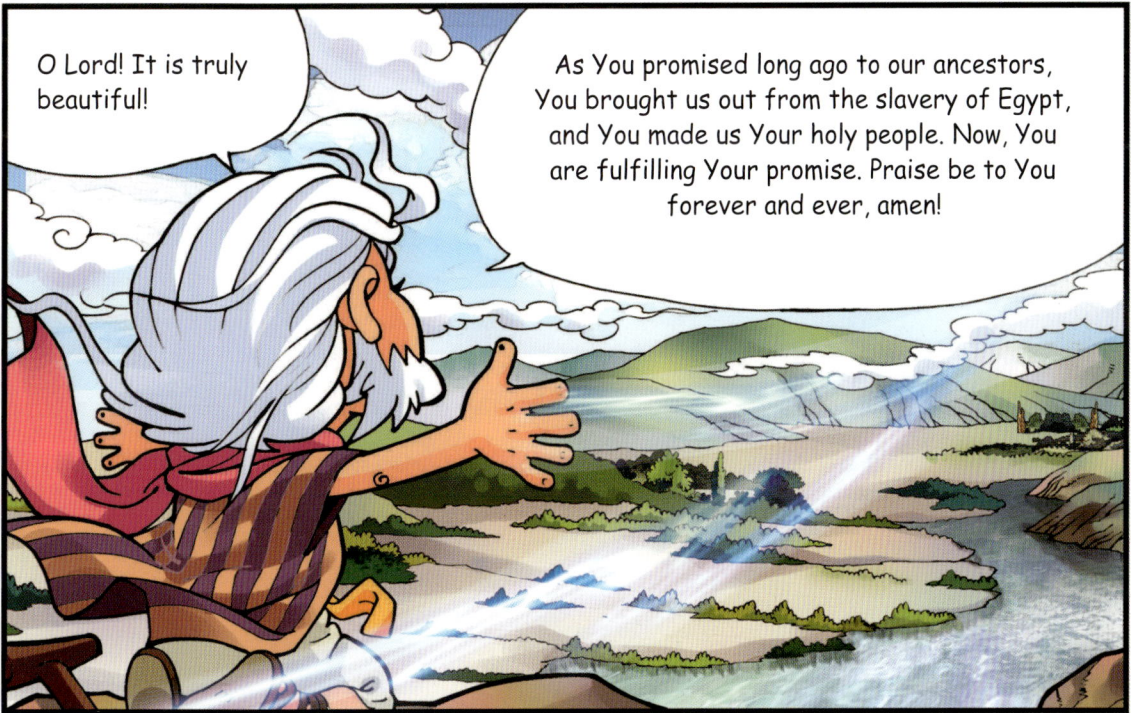

O Lord! It is truly beautiful!

As You promised long ago to our ancestors, You brought us out from the slavery of Egypt, and You made us Your holy people. Now, You are fulfilling Your promise. Praise be to You forever and ever, amen!

MOSES SPOKE DIRECTLY WITH GOD, AND HE TAUGHT THE PEOPLE HOW TO BE A HOLY PEOPLE OF THE LORD. MOSES WAS A GREAT LEADER WHO LED HIS PEOPLE OUT OF SLAVERY IN EGYPT. HE WAS ABLE TO SEE THE PROMISED LAND FROM MOUNT NEBO. THEN, HE PASSED AWAY AT THE AGE OF 120.

End of Power Bible #2. Please stay tuned for Power Bible #3!

BIBLE STORIES TO IMPART WISDOM

POWER BIBLE **2** Moses, Leader of the Israelites

Published by: Green Egg Media, a division of Three Sixteen Publishing Inc.
316publishing.com

Copyright © 2005 by Kim Shin-joong
English translation: Copyright © 2011 Green Egg Media
Reprint 2022
Original published in Korea by Mirae N Co., Ltd. (I-seum)
This translated edition was published by arrangement with Mirae N Co., Ltd. (I-seum)

The names of people and places, Scripture quotations, and passages in this book have been written in accordance with the New American Standard Bible (NASB®).
"Scripture quotations taken from the New American Standard Bible®
Copyright © 1960, 1962, 1963, 1968, 1971, 1972, 1973, 1975, 1977, 1995 by The Lockman Foundation
Used by permission." (Lockman.org)

Acknowledgments:

Green Egg Media, Inc. would like to acknowledge the following individuals for their commitment and contributions: Sunny Han, Jenny Eunsook Kim, Herbert Yu, Paul Lee, Allison McMillan-Lee, Christine Lee, Cindy Kim, Swan Kim, John Kim, Gary Low, Rex Morishita, and Greg Joe.
Special thanks to Gene Yang, award-winning cartoonist, for his help and insight about comics in education.

PB0002
ISBN 978-937212-01-8
Printed in Korea by Codra Enterprises, Inc.   codra.com

Green Egg Media   a division of   **THREE SIXTEEN PUBLISHING**

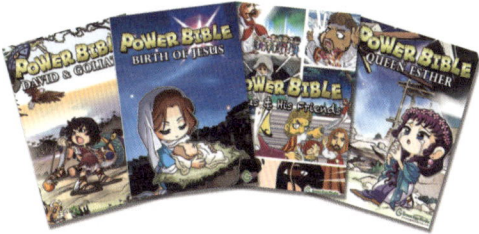